# cross|ing

**Time.Space.Movement** OLU OGUIBE

with contribution by Okwui Enwezor

CONTEMPORARY ART MUSEUM, UNIVERSITY OF SOUTH FLORIDA

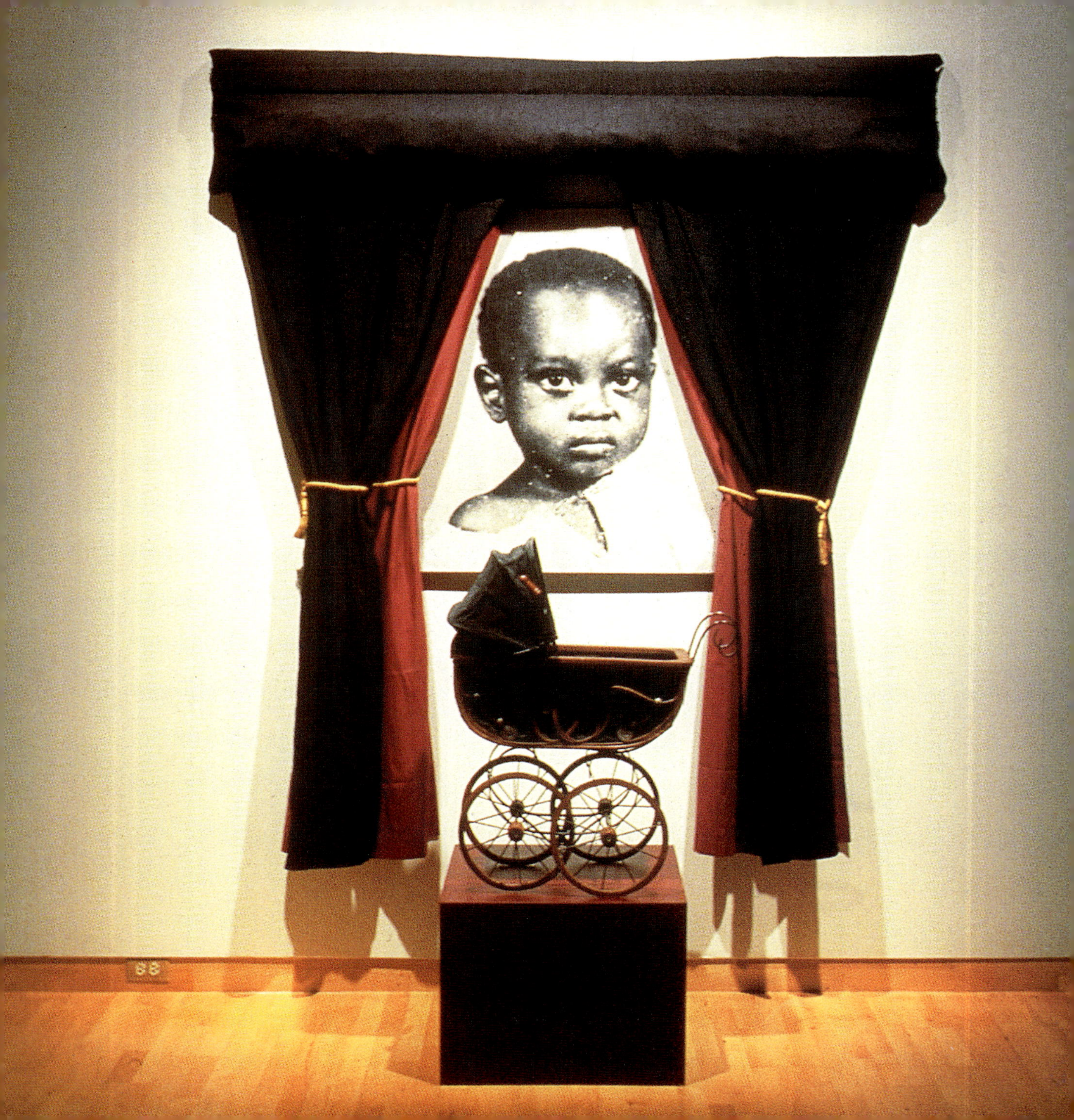

Time.Space.Movement

# cross|ing

Oladélé Bamgboyé
Bili Bidjocka
Gordon Bleach
Kendell Geers
Lubaina Himid
Marcia Kure
Houria Niati
Olu Oguibe
Tracey Rose
Folake Shoga

"Cross/ing: Time.Space.Movement"
Contemporary Art Museum
University of South Florida
September 4–October 18, 1997

Track 16 Gallery
Santa Monica, California
February 28–April 24, 1998

Indianapolis Museum of Art
Indianapolis, Indiana
June–August, 1999

USF Contemporary Art Museum

Margaret A. Miller, Director
Alexa Favata, Assistant Director
Vincent Ahern, Coordinator, Public Art
Peter Foe, Curator of Collections
Tony Wong Palms, Exhibitions Supervisor
Don Michael Fuller, Preparator
Larry Mize, Preparator
David Nickels, Public Art Assistant
David Waterman, Security
Randall West, Accountant
Lissie Cosme, Program Assistant
Kristen Welch, Program Assistant
Brian MacElhose, Program Assistant
Kim McDaniels, Events Coordinator

Editor: Melissa Ho
Book Design: Marika van Adelsberg
Copy Editor: Sherri Schottlaender
Photography: Peter Foe
Photography of *Memorial to a Youth:* Olu Oguibe

Frontise: Olu Oguibe, *Buggy (For an Unknown Child)*, 1997


Published by Smart Art Press
2525 Michigan Avenue, Building C1
Santa Monica, CA 90404
310.264.4678 TEL
310.264.4682 FAX
www.track16.com

ISBN# 1-889195-24-3
Smart Art Press Vol. 5, No. 43

Project assistance for "Cross/ing" provided by:
The Rockefeller Foundation
and
The Elizabeth Firestone Graham Foundation

Printed and bound by Jomagar, S.A., Spain

The USF Contemporary Art Museum is recognized as a major cultural institution by the State of Florida and received funding through the Florida Department of State, the Florida Arts Council, and the Division of Cultural Affairs. The USF CAM is accredited by the American Association of Museums.

Time.Space.Movement

# crossing

## Contents


6 **Preface**

8 **Acknowledgments**

10 **crossing Time.Space.Movement**
OLU OGUIBE

14 Oladélé Bamgboyé
18 Bili Bidjocka
22 Gordon Bleach
26 Kendell Geers
30 Lubaina Himid
34 Marcia Kure
38 Houria Niati
44 Olu Oguibe
48 Tracey Rose
52 Folake Shoga

57 **Between Localism and Worldliness**
OKWUI ENWEZOR

64 **Artist Biographies**

66 **Checklist**

69 **Selected Bibliography**

DR. JOHN L. SMITH

Dean, College of Fine Arts
University of South Florida

# Preface

The University of South Florida's Stuart S. Golding Endowed Chair in African Art, and the Center for Africa and the Diaspora of the Institute on Black Life inspired the Contemporary Art Museum to develop an exhibition of contemporary art from Africa. With the recommendation of Grace Stanislaus, Director of the Museum for African Art in New York, Margaret A. Miller, Director of USF's Contemporary Art Museum, commissioned Dr. Olu Oguibe to design an innovative exhibition to present the current dynamic of the African diaspora. The exhibition was shared with the Museum of African American Art located in downtown Tampa. In 1996 Dr. Oguibe, a Nigerian-born artist, curator, and art critic, accepted an appointment as Visiting Associate in African Art and Stuart S. Golding Scholar in the USF Art Department. Dr. Oguibe designed "Cross/ing: Time.Space.Movement" in consultation with Mr. Okwui Enwezor, a Nigerian-born, New York-based independent curator, art critic, and the Artistic Director of the "2nd Johannesburg Biennale," and Mr. Carl Hazlewood, the cofounder of the Aljira Center for Contemporary Art in Newark, New Jersey. The organization of the project was coordinated by Alexa Favata, Assistant Director of the USF Contemporary Art Museum.

"Cross/ing" introduces the work of ten leading artists: Oladélé Bamgboyé, Bili Bidjocka, Gordon Bleach, Kendell Geers, Lubaina Himid, Marcia Kure, Houria Niati, Olu Oguibe, Tracey Rose, and Folake Shoga. The artists, born in the African countries of Nigeria, Zimbabwe, Cameroon, South Africa, Tanzania, and Algeria, explore some of the most compelling issues of our time, including the globalization of culture, immigration, migrancy, foreignness, identity, and advances in technology.

The general impression that Africa is culturally isolated is dispelled by "Cross/ing." All but two of the artists represented in the exhibition—Marcia Kure and Kendell Geers—currently live outside of Africa, in England, France, or the United States. The artistic forms embraced by these artists demonstrate a rigorous engagement with the artistic debates of the international art community. The conditions of artists living and working at the end of the twentieth century have been transformed by the electronic communication revolution; the works in this exhibition allow for the examination of the forces that formulate culture, and "Cross/ing" provides a fresh understanding of how we encounter, assimiliate, disseminate, translate, revise, and understand diverse cultural values.

The University of South Florida campus and the Tampa Bay Community benefited from the participation of Okwui Enwezor and Carl Hazlewood in a symposium presented in conjunction with the opening of the exhibition. The residencies of Kendell Geers, Houria Niati, and Marcia Kure provided students, faculty, and the broader community with an opportunity to understand more fully the particular positions offered by their artistic production.

Congratulations to Olu Oguibe for his groundbreaking exhibition, and to Margaret Miller and Alexa Favata for their persistent and skilled efforts in developing this important project for the University of South Florida. General assistance provided by the Rockefeller Foundation and the Elizabeth Firestone Graham Foundation made the project and catalogue possible. We are pleased to be able to travel "Cross/ing" so that other communities may be introduced to this exceptional emerging group of African artists.

MARGARET A. MILLER

Director and Professor
USF Contemporary Art Museum

# Acknowledgments

Our perceptions of our place in a global culture have been transformed by advanced technologies for communication, as well as immigration and mass migration. Geopolitically we must consider the meanings of national affiliation, changing economies, urbanization, and the effects of colliding cultural values. "Cross/ing: Time.Space.Movement," an exhibition of African contemporary artists, curated by Dr. Olu Oguibe, has been an inspiring challenge from inception to realization. Dr. Juel Smith, Director of the Institute on Black Life, and Dr. John Smith, Dean of the College of Fine Arts, believed that it was not only important but possible for the University of South Florida Contemporary Art Museum to organize an exhibition of contemporary art from Africa. Roberta Golding made it possible to undertake a project of this magnitude when she established the Stuart S. Golding Endowed Chair in African Art to honor her husband and to encourage the study and promotion of the rich artistic resources from Africa which underpin so much of American culture. "Cross/ing" examines the forces that formulate culture.

Research for the project was initiated more than three years ago by Corine Gaile, a faculty member in the Art Department, and by Grace Stanislaus, now director of the Museum for African Art in New York. Their research trips to Africa and their knowledge in this field ultimately led to Dr. Olu Oguibe; as an artist, teacher, critical writer, and curator born in Nigeria and educated in Nigeria and England, Dr. Oguibe was fully prepared to curate an exhibition from an African perspective. He actively participates with a network of other scholars—including the consultants for this project, Okwui Enwezor and Carl Hazlewood—to introduce a generation of artists who were born on the African continent and are contributing to the global discourse on the neo-avant-garde.

"Cross/ing" is designed to provoke the viewer

to consider the crossing of intellectual, geographic, ideological, psychological, and stylistic boundaries. Systems of iconography and definitions of representation contain both perceptual and aesthetic conventions that are informed by a complex layering of information. Several of the ten artists represented in the exhibition directly (and indirectly) address issues of ethnocentrism, bias, stereotyping, prejudice, discrimination, and racism.

Alexa Favata, Assistant Director of the USF Contemporary Art Museum, provided the skilled leadership in organizing this ambitious project, and she was assisted by Lissie Cosme. Tony Palms, USF CAM's Exhibitions Supervisor, and his staff, Larry Mize, Don Fuller, and Brian MacElhose, worked directly with the artists to realize the installation designed for this exhibition. I am grateful to Tom Patchett and Smart Art Press for publishing the catalogue, and to Smart Art Press editors Melissa Ho and Pilar Perez for their work on the book. Marika van Adelsberg should be acknowledged for her brilliant design, and Sherri Schottlaender for her meticulous copyediting. Generous grants from the Rockefeller Foundation and the Elizabeth Graham Firestone Foundation provided critical financial support for the project. Kendell Geers's participation in "Cross/ing" was made possible in part by the South African Department of Science and Technology, Arts and Culture.

"Cross/ing" inaugurates an ongoing commitment by the USF Contemporary Art Museum to develop exhibitions and education programs that explore the African diaspora and its contemporary manifestations.

OLU OGUIBE

crossing

# Time.Space.Movement

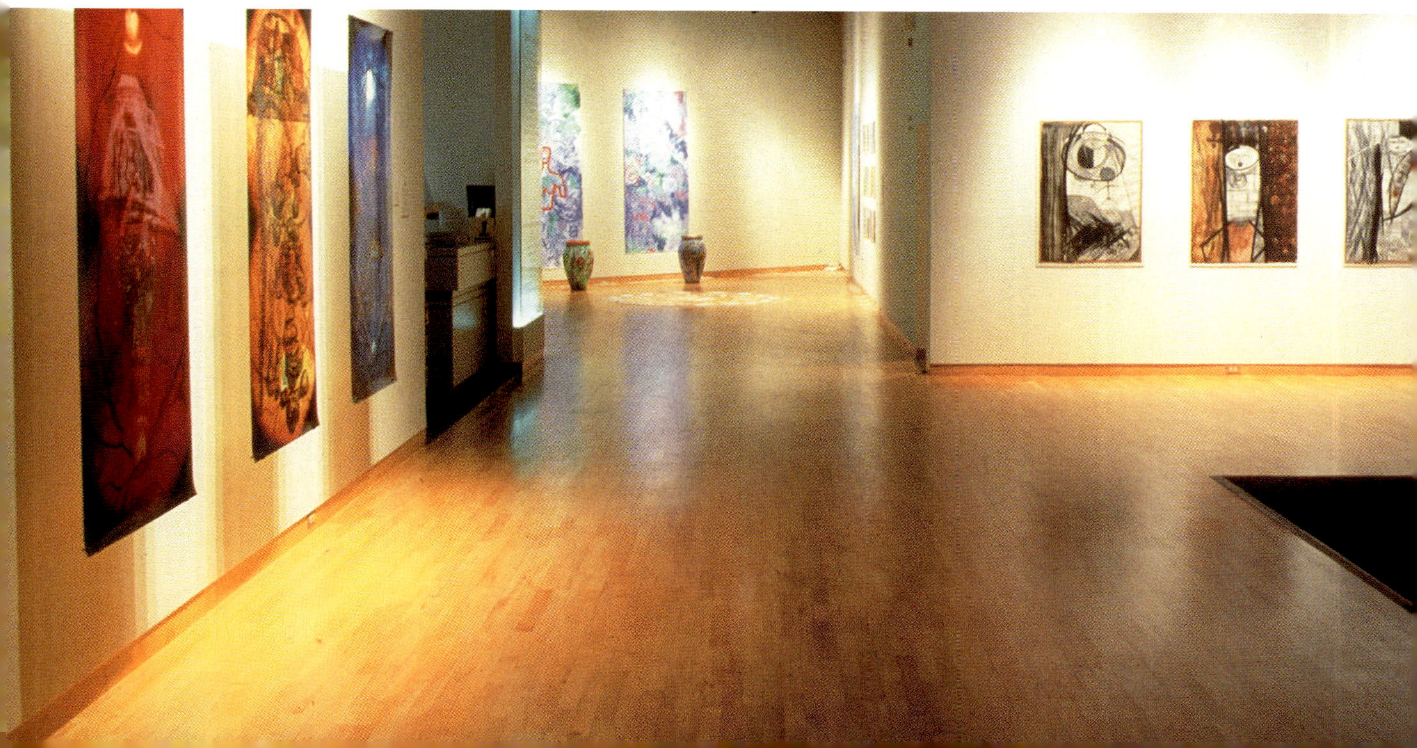

At no moment in history has the truth been clearer than it is today that movement is the human condition. Whether manifested in the geophysical, demographic flux constantly exacerbated by the prevarications of political or economic forces, or in the utilization of the hitherto unimaginable possibilities of information technology and the mechanics of virtuality, movement, transition, transformation, relocation, transfiguration, and subjection to the enormous fluidities of location and identity are the imperatives of existence in our age. Today, to survive is to be constantly on the move, to engage in an endless shift, to recognize and acknowledge the underlying fluidity in the nature of things.

Where fictions of certainty once resided there is a growing awareness of uncertainty, and whatever was once taken for granted, be it a sense of self or a sense of home, no longer offers the comfort of certitude. To be is to continually engage in the vacation and reoccupation of moments and localities into which we are often forced, and it is this constant transition that defines the state of ideas and practices. In art as in the perception and articulation of material culture, nothing survives that still suffers the vulnerability of fixity.

The material and ideological geographies that we occupy have become what the critic Octavio Zaya refers to as interzones, localities where "transgressions, differentiations, dislocations, and transfigurations . . . challenge the sense of integrity we relate to culture."[1] Beginning with the collapse of colonial empires and the consequent reversal of migratory patterns between empires and colonies, no territory escaped reconfiguration, no boundary remained unbroachable.

In the case of Africa, in the past three decades a great mass of its peoples has crossed borders into other localities and sites, extending the boundaries of the continent beyond its previous geographical limits and thereby

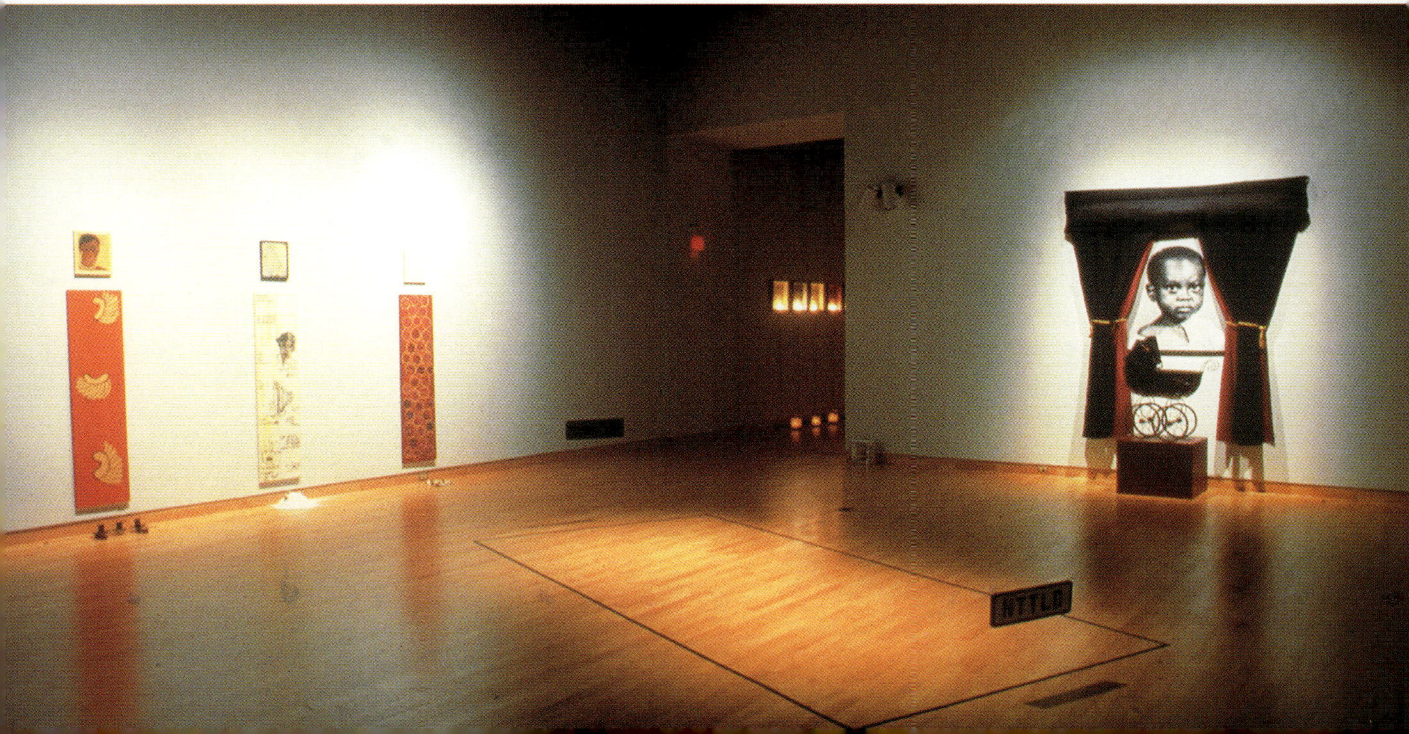

reconfiguring the precepts and canons of its cartographic definition. Africa is no longer the landmass between continents; it has become a referent for many who reside within multiple other borders, those who carry the continent but not the landmass with them. For many, even, Africa has become what the author Salman Rushdie refers to as an "imaginary homeland."[2]

The question of identities which arises from this defiance and defilement of borders is one of great interest and is a source of anxiety for those who look to Africa for a fictive essence. And perhaps understandably so for a continent named (and for long largely defined) from the outside, a land of cultures wishfully debarred from the natural alterity of the rest of humanity. Yet fluidity and mutability were always the defining elements in the material culture and world view of African cultures, reflected in a widely used West African aphorism: *No condition is permanent*. Thus is the continent able to transform and transfigure in order to accommodate its myriad inclinations and ethnicities, from Arab African to Caucasian African, from Bantu to Asian African. Thus are the cultures and polities able to contain the variegated fortunes and destinies that are the hallmarks of Africa's uniqueness.

Not only do African artists reflect the interzonality of our millennial age in their work, they also live it through the internationality of their practices and domiciles. While a good number live and practice on the continent, a great many others (like other Africans irrespective of their backgrounds and calling) live and pursue their preoccupations on the same stages as other peoples and races from around the globe who today live and work in the numerous metropolises of the world. From South Africa to Germany to the United Arab Emirates to the United States of America, African artists are constantly crossing: stepping through border posts and entry stations, through guarded spaces and sites, over the boundaries of critical and spectacular attention, into the supposedly exclusive sites of contemporary art practice, engaging in the same questions and issues as their contemporaries, bringing to this engagement their experience of the flux that defines our age.

Time, therefore, and locality, and movement in time (whether chronological or zonal)—these define the art of the moment. Location and dislocation, situation and misplacement, accommodation and alienation, attachment and disenchantment, affection and disaffection, but most importantly, the transfigural interstices between these, point up an art different from that manufactured in the imaginations of brokers of pristine quintessence. In their continual defiance and defilement of borders, and of the cross that the outsider imagination places upon them, contemporary African artists concretize the aphorism that nothing is, or ought to be, the way it is thought to be.

"Cross/ing" presents the work of African artists who, though no longer bounded by the old affiliations of geography and race, inevitably reflect (on) their common claim to Africa in their work. It features the work of five female and five male artists from the north of the continent to the south, across racial divides, artists who by the nature of their practice and cultural milieu belong in a larger world and divide their time between Africa and other centers of contemporary cultural practice, artists who traverse the numerous geographies of contemporary art and reality, whether physical, conceptual, or metaphysical.

# OLADÉLÉ BAMGBOYÉ

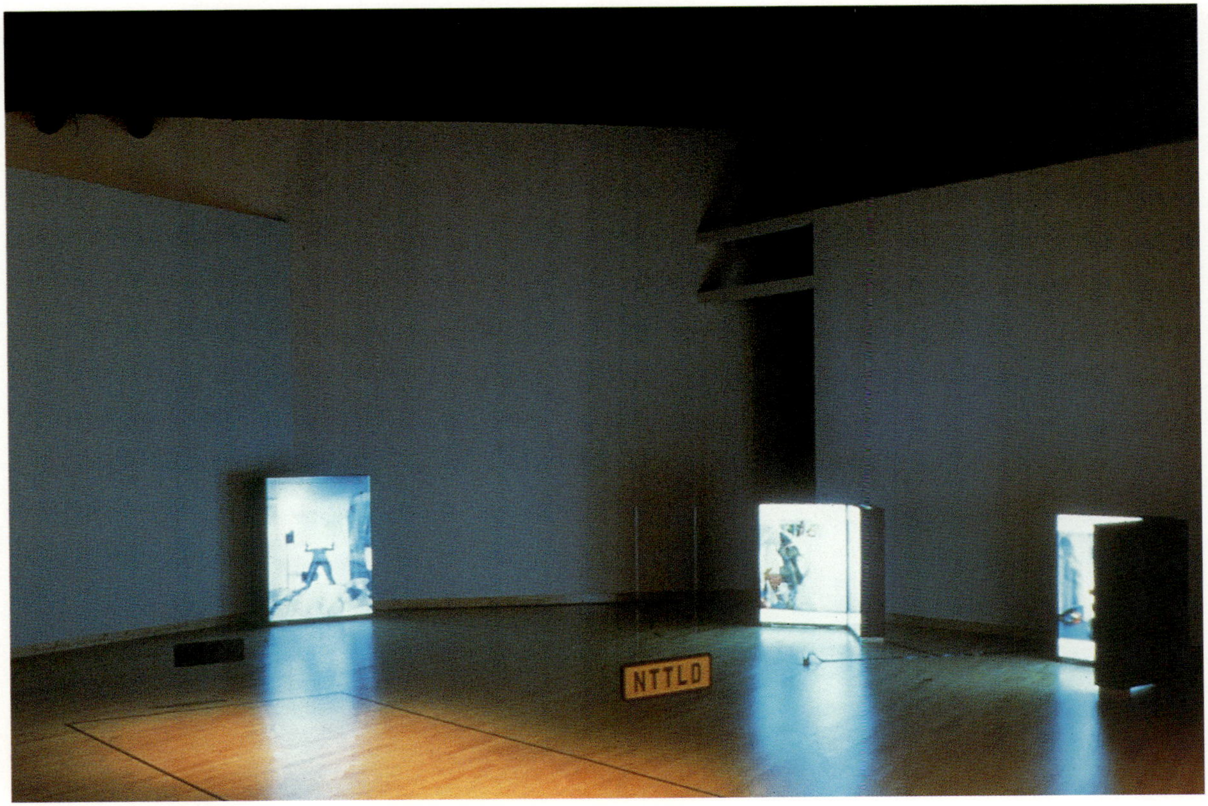


Oladélé Bamgboyé
*left to right:*
Panel #4, from the series "Das Lichthaus"
1989
Images 1, 3, 5, 6 from the series "Celebrate"
1994

Oladélé Bamgboyé was born in Benin, Nigeria, and grew up in Scotland. Crossing over from chemical engineering to photography, Bamgboyé's early work consists of highly charged black-and-white compositions that deal with the oft-charged question of the invention of the black body, especially in western society. Bamgboyé's early work is both an appreciation of the realities of the black male body and an aesthetic celebration in the tradition of Geoffrey Holder and Bamgboyé's own precursor, the photo-artist Rotimi Fani-Kayode, both of whom influenced his early work. It is also an informed and determined critique of the distortive constructs of that body, images whose purpose is to bring the voyeurism that manufactures such constructs into crisis.

Growing up as a minority away from his society of origin, Bamgboyé had both a historical awareness and personal experience of the gaze that turns the black male body into an abject object of desire and wonder, an object of mystery, trepidation, and lust. In time he gradually discovered his own unintended availability as a body: as a young, strapping, black male who wore his hair in locks, he had become the unwitting quintessence of a type, a composite of desirable parts and an unsuspecting victim of a tradition within which the simple humanness that he could take for granted in his society of origin is infringed upon and essentially compromised. Bamgboyé, like others of his kind living and growing up in the civilized mildness of European prejudice, now bore the cross of a blackness of which he had no awareness in Africa, a cross the burden of which, as James Baldwin wrote, *is always upon us.* At one station of this cross, Bamgboyé cut his hair, removing the locks that were beloved of those who see the black male body before they see its humanity. In a reflective video piece that deals with this experience, *The Hair or the Man?,* he reproduces in text a question that summarized the inescapability of the experience and the futility of his sacrifice, a question that every young, black male at that station of the cross must suffer time and again: Why did you cut those sexy locks? It is a question which, though oft

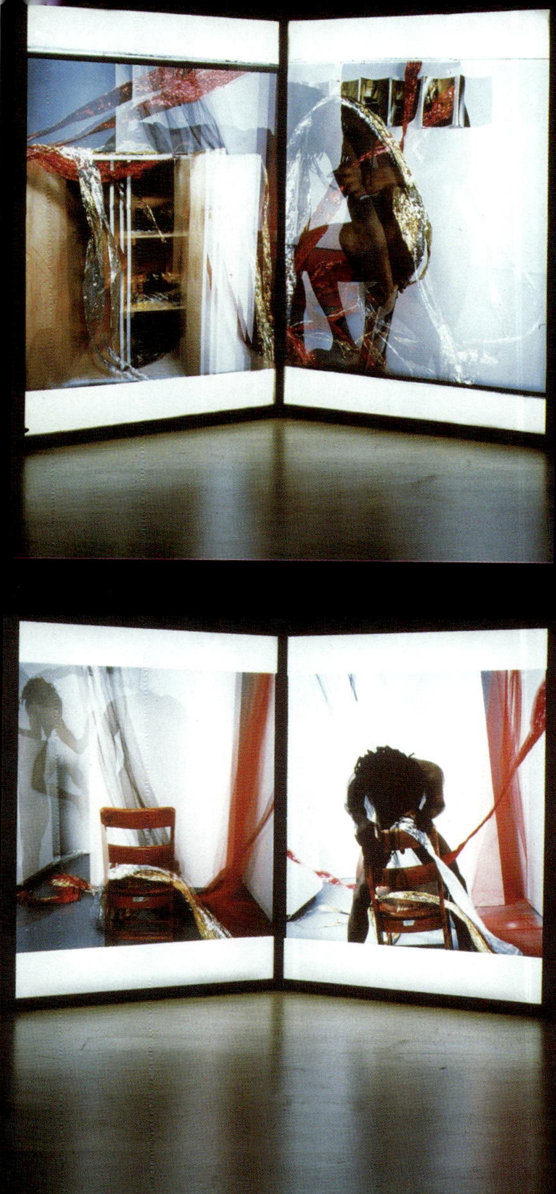

Oladélé Bamgboyé
Images 1, 3, 5, 6 from the series "Celebrate"
1994

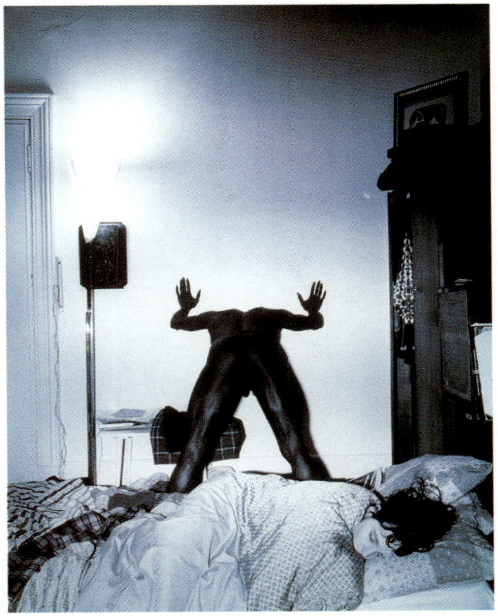

Oladélé Bamgboyé
Panel #4, from the series "Das Lichthaus"
1989

intended as compliment, is nevertheless rooted in the very dark depths of prejudice and reductive bigotry that perceives of the addressed not as man but as curio. In cutting his hair Bamgboyé had been made to pay a price, to give something of himself, to shed a part of his being in the struggle to reduce the burden of the cross of blackness, and to survive. As Maya Angelou once wrote, he had allowed—had been made to watch—another little portion of his territory taken away. In crossing he had come upon a territory which, in threatening his sense of wholeness, nevertheless turns him into a man who questions.

Bamgboyé's photographic tableau from the series "Das Lichthaus," produced in Germany in 1989, is a startling commentary on the question of race and representation, as well as on perception and interpretation. In a black-and-white photograph, a nude black male faces away from us, splayed out against a wall in a position of innumerable resonances that at once reminds us as much of the humiliation of the body search—a constant of the black male experience in the West—as it does of the defiant gesture of a face-off. In the foreground a young white female lies asleep in a well-made bed, horizontal to the picture plane, unruffled and apparently oblivious of the ritual of postures behind her, serene in her sleep. She appears both complicit in the indeterminate narrative of the picture and in contrast to the turmoil of place and history that her male companion stirs in us. We ponder and contemplate her place in this narrative, in the immediate and potent environs of her location, and in the history of the cross-cultural gestures that bring both figures together. It is an explosive image: the artist's careful excision of the male figure's head, the absence of the black male face and the ruthless presence of the black body, contrasted with the fullness of the white female countenance on a concealed body, set us off on multiple paths of reading, meaning, insinuation, and emotion. The daintiness of the white female figure, carefully protected from our gaze, concedes wholeness to her and invests her with innocence and a naiveté that draws our empathy, even concern for her safety. On the other hand, the deliberately confrontational posture of the black male figure and his nakedness challenge our direct gaze even as his partial erasure speaks to a denial of wholeness.

In *Celebrate* (1994), Bamgboyé presents a seductive series of composite images that trace the transit of a figure through a terrain of polythene distractions and constraints to the moment of extrication, suggesting the ecstasy of a breakthrough, a final moment of self-discovery after a labyrinthian sojourn. We are reminded of the protagonist in the poet Christopher Okigbo's "Distances," weaving his way homeward, until in final rapture he declares:

**. . . in the orangery of immense corridors**
**I wash my feet in your pure head, O maid**
**I have entered your bridal chamber; and lo,**
**I am the sole witness to my homecoming.**[3]

BILI BIDJOCKA

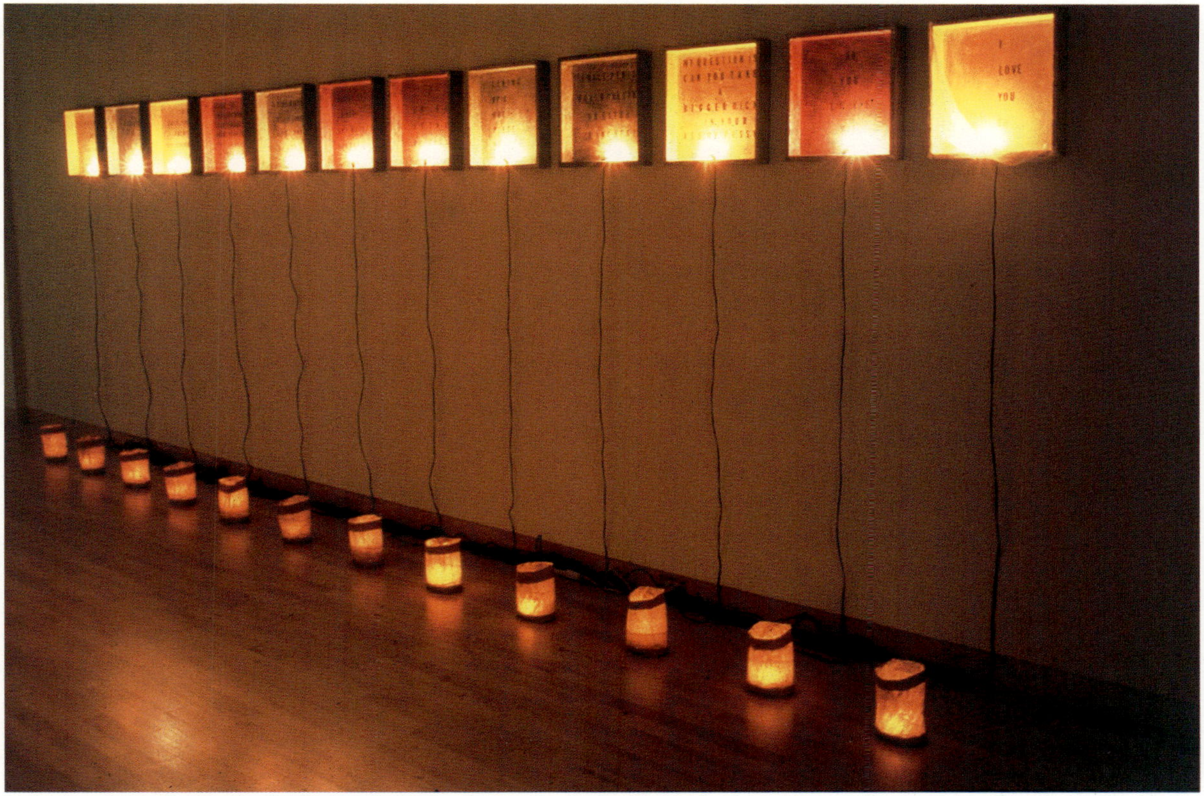


Bili Bidjocka
*Explicit Lyrics*
1995

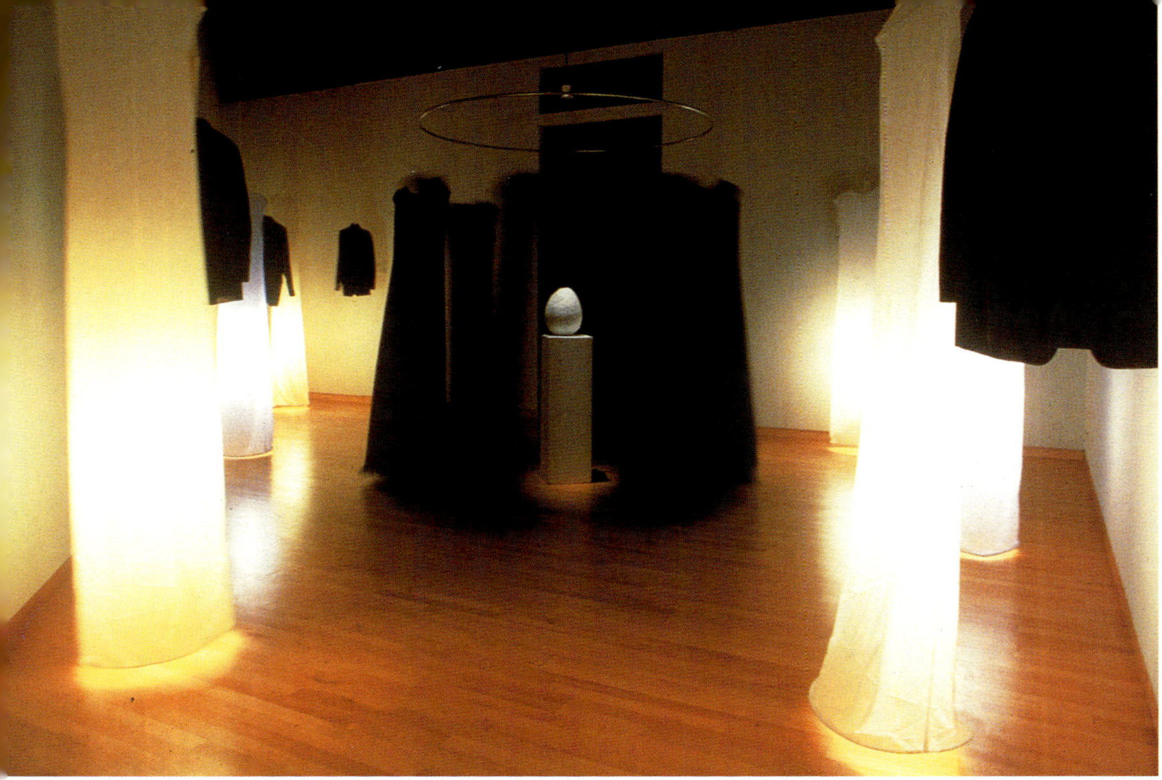

Bili Bidjocka
*Untitled (Witches' Ball)*
1994

Bili Bidjocka was born in Bidjocka, near Douala, Cameroon, and was taken to France by his parents at the age of twelve. There he received the rest of his education and began to practice as an artist, returning occasionally to the town of his birth in West Africa but living in and perceiving himself as part of the Paris of his upbringing and location as an artist, albeit a part constantly wedged into the slippery terrains of an uncertain belonging. As a conceptual artist Bidjocka is primarily preoccupied with the thought patterns and ideas he wrestles with as he translates them into visual form, his focus being the terrors and thrills of that wrestling, that tussle with the incorporeal as he invests it with form.

Bidjocka's *Explicit Lyrics* is truly about crossing: it is about the innocence and complicity of the stranger in a new city, about the pitfalls of language and the seductiveness of the occluded. It is also about the fact that, quite often, our introduction to an unfamiliar culture is through the route of the rude. In 1995 Bidjocka and his friend and fellow Paris artist Bato Ndenga were invited to New York by friends to work in an independent artists' space. Both were faced with the self-imposed challenge of producing a show at short notice using the city as object. For his installations, Bidjocka began by gathering clothing from friends and strangers. Then he stumbled upon the personal services media in New York and became fascinated by its language, which, as a French speaker with little English, he found most intriguing and therefore, as is often the case with the intriguing, most seductive. He was able to locate lyricism in a language that, evacuated of meaning and social associations as it was for him, became text as curio. And even as meaning was returned to it through translation, the inevitably cynical nature of the text and its source industry froze it in the frame of the exotic. It became fetish.

Such keen, almost tongue-in-cheek fascination with text and the allures and failures of language comes as no surprise to us when we position it within the tradition of French semiotic preoccupation; the appeal of language

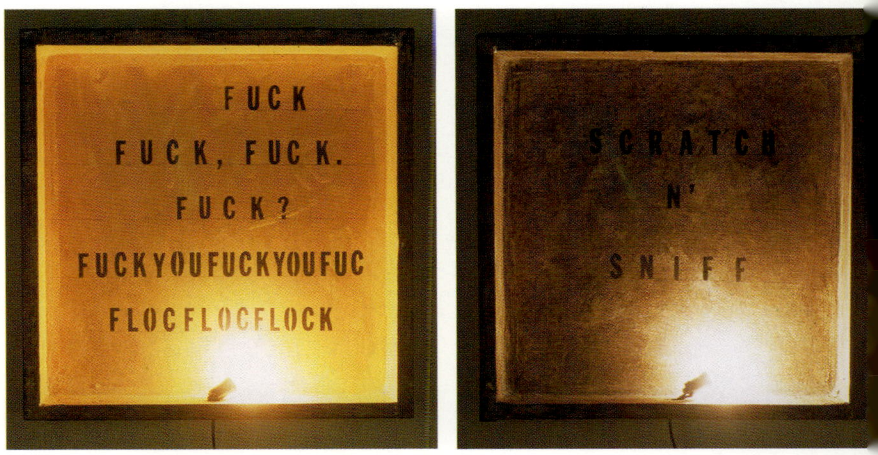


Bili Bidjocka
*Explicit Lyrics* (detail)
1995
mixed-media installation

is strong for a man coming from a culture that savors language and the pleasures of the text. Nor would it fail to make sense when we consider Bidjocka's other heritage, in Africa, where language holds the same pleasurable appeal as it does in France. The pleasure, and the emphasis, therefore, reside not in the explicitness of the phrases as for those who have immediate access to the language, but rather in their lyricism, their duplicity as signs, in their ability to resonate on a different pitch once devoid of the constraints of meaning, perceived from *across* and *beyond* the boundaries of literal signification and the legislature of social investment.

In *Untitled (Witches' Ball)* (1994), Bidjocka deals with absence and evacuation, as well as with the resilient occupation of spaces which is characteristic of the human spirit. In this installation piece, empty high-society ball costumes float sensuously across the room to the sound of a gentle waltz, residual signs of a prior presence and a once-animated joie de vivre. We are reminded of the ghost houses of Laurenco Marques, or of Gabriel Garciá Márquez's deserted cities where the spirits of the dead linger, laying continuous claim to moment and territory, reminding us that there is no *terra nullus,* no space uninhabited by the residues of our passing. Around the room the ghosts of the nobility hang around in little groups, each lady to her gentleman, represented by their empty jackets and slinky dresses, the wineglasses now resident only in our imagination, and so too the highly polished shoes and silk scarfs and powdered faces, the mock grins of civility and the hidden furnaces of illicit desire, the repressions and follies, the airs of accomplishment and contentment, of snobbery and injured pride. The music plays on, but all else is gone now, our only reminder the emptiness and aloneness of this uninhabited apparel that bears witness still to its owners and the lives they led. There is a silent poetry in the piece, and a touch of sadness as the artist, though he makes a case for the resilience of human presence, nevertheless points to the fact of our mortality.

GORDON BLEACH

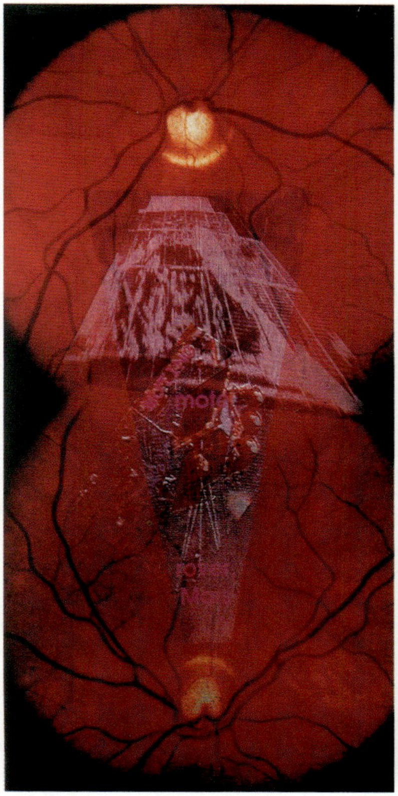


Gordon Bleach
*Red Mote: rover, Mars* from
"Double Vision: The Mote Series"
1997

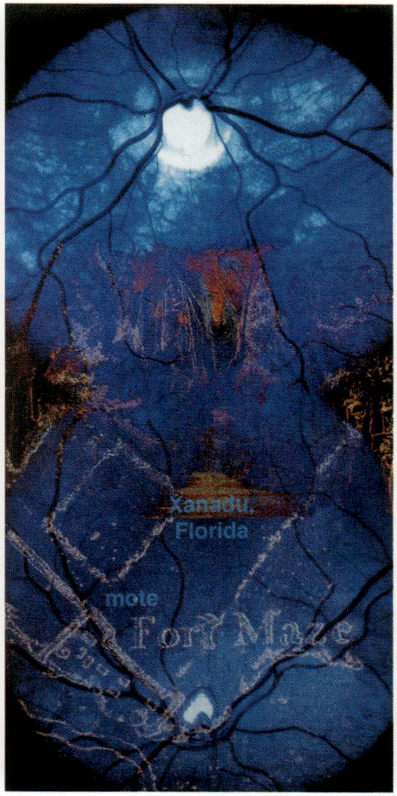


Gordon Bleach
*Blue Mote: xanadu, Florida* from
"Double Vision: The Mote Series"
1997

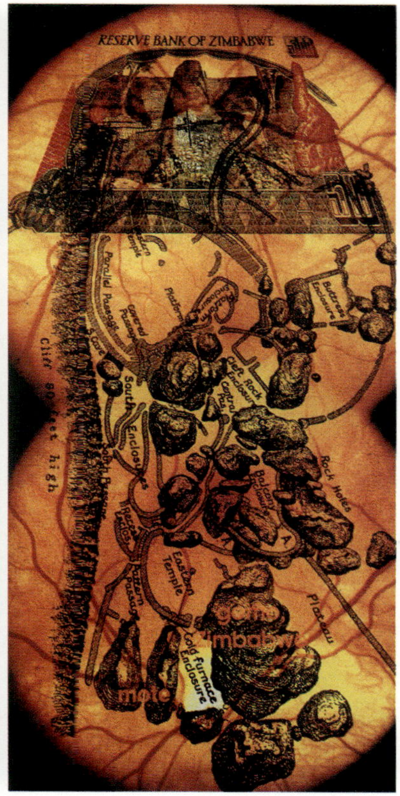


Gordon Bleach
*Gold Mote: gomo, Zimbabwe* from
"Double Vision: The Mote Series"
1997

Gordon Bleach was born in what is today Zimbabwe; he is descended from three centuries of Caucasian Africans, his maternal ancestors having settled in Southern Africa in the eighteenth century. As a youth Bleach fled his country, then still under the name of the colonialist Cecil Rhodes, to avoid conscription into its army as it waged a war to preserve white minority rule. After years of study as a mathematician in Cape Town, South Africa, he moved to America where he began to pursue fully his lifelong passion for art rather than the sciences. In between, he traveled round the world, as if to rid his psyche of the trauma of flight by discovering movement as pleasurable quest. He also began to make art that deals with geographies and their relationship to inscribing or eliciting entities and notions, their virtuality outside of the perceptive parameters of the body, their subordination to the overarching reaches of the imagination, their hold on memory and recollection.

Bleach works on the ineluctable fluidities of territories, the notion of space, identity, and history as an elaborate mesh with a messy edge where ideas and places and moments interweave without apparent pattern yet with an inherent logic, making absolute placement a futile preoccupation. In his "Mote" series he uses digital technology to manipulate and translate vistas and memories into aesthetic and conceptual objects. The central motif—the mote—is the bright area of the retina (also known as the blind spot) which collects information and transmits it to the brain for cognition. He uses images of his own retina as a frame into which other signifiers fall. In *Red Mote: rover, Mars,* he makes reference to the Mars Rover, the little robot that was landed on the surface of the planet Mars in 1997 as a mote, a tool for the collection and collation of information on what he describes as "the red, canal-veined planet of energy and war." In *Gold Mote: gomo, Zimbabwe,* the reference is to not only his country of birth and to the Shona word for mountain—*gomo*—but also to the practice of gold-panning, which is characteristic of a certain moment and location in the country's history and a sign for the larger complexity of currency flows and their politics. In *Blue Mote: xanadu, Florida,* Dr. Bleach reconnects the Xanadu of Coleridge's Kubla Khan to its physical referent, Florida,[4] region of the Fountain of Youth and shore to the mysterious Bermuda Triangle. This way the artist leads us back to the essential vulnerability of memory as it goes through loss and resurrection.

Bleach's use of digital technology is in itself a manner of crossing: a subversion of certainties and so-called objectivities, a reiteration of the essential shiftiness of reality. Digitization and pixilation return us to the fundamentals of our nature, to the looseness of our constitution as conglomerations of free and ever-shifting particles, as bundles of moving entities, as collations of disparate multiples rather than monolithic entities. By moving and manipulating the tiny digits and pixels that make up his images, the artist creates an analog to our own selves and ridicules our fictions of purity and fundamentality in a world so intricately interwoven that none of us is an entity free and unto itself. Digitization speaks to identity and to the fact that, at the most basic and fundamental level, all reality is composed of the same particles and none of us is, in essence, *particularly* different from the other.

In *Slcsh* (1997), an architectural pattern of images and objects, Bleach reconstructs both the slash in the word *"cross/ing"* and the idea of the crossed boundary, as well as the spuriousness of permanence. In a project produced in the early 1990s,[5] the artist chose a construction site in upstate New York where every night he would insert himself into the ghostly architecture of the site, and with a Mag-Lite in hand, as if writing in the dark, he would stand for hours tracing the fictional contours of imaginary architectures until they registered on his photographic plate. Then he would disappear, having made his body unavailable as he stood in the dark beyond the reach of the photographic register. Eventually he created virtual cities made up entirely of light and of the drawn line, cities that exist only through inscription, through the coaxing power of

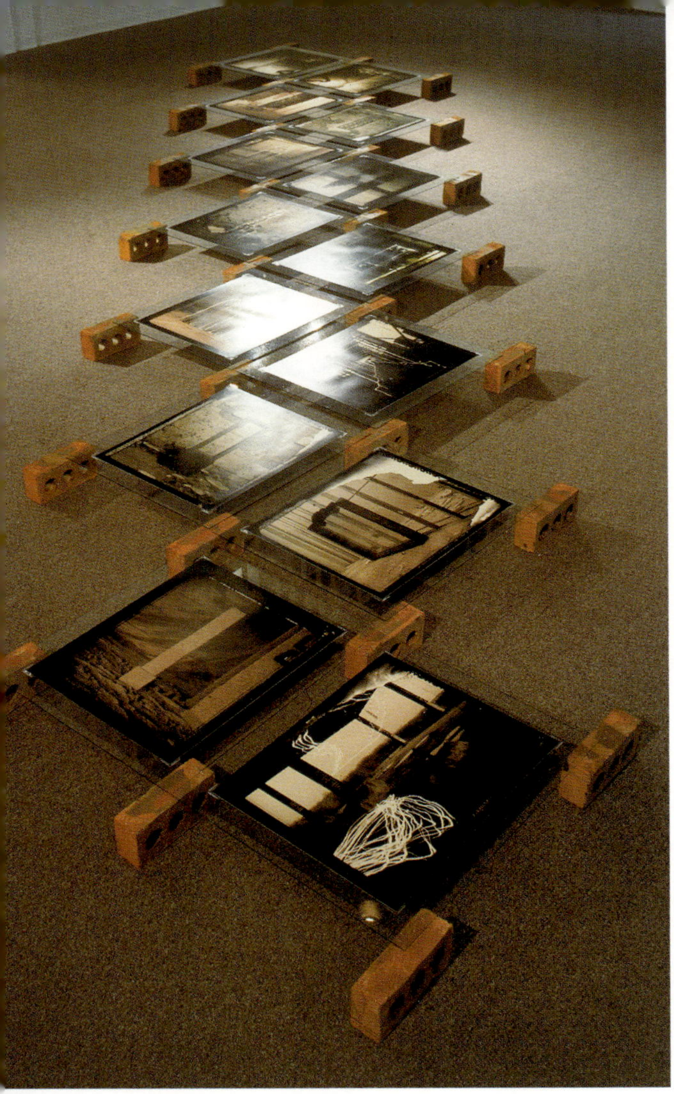

Gordon Bleach
*Slash*
1997

the sign. In the same vein he reaffirmed the death of the author, for though he was the only real object in the camera's frame, he was able to elude that frame and instead fill it with the virtual, the unreal, the temporary and transitory, that which moves until it achieves permanence. With these images simulated and constructed through the make-believe processes of drawing and virtual positioning, Bleach crisscrosses and cancels out at the same time, creating both a structure and the discourses of its untenability.

KENDELL GEERS

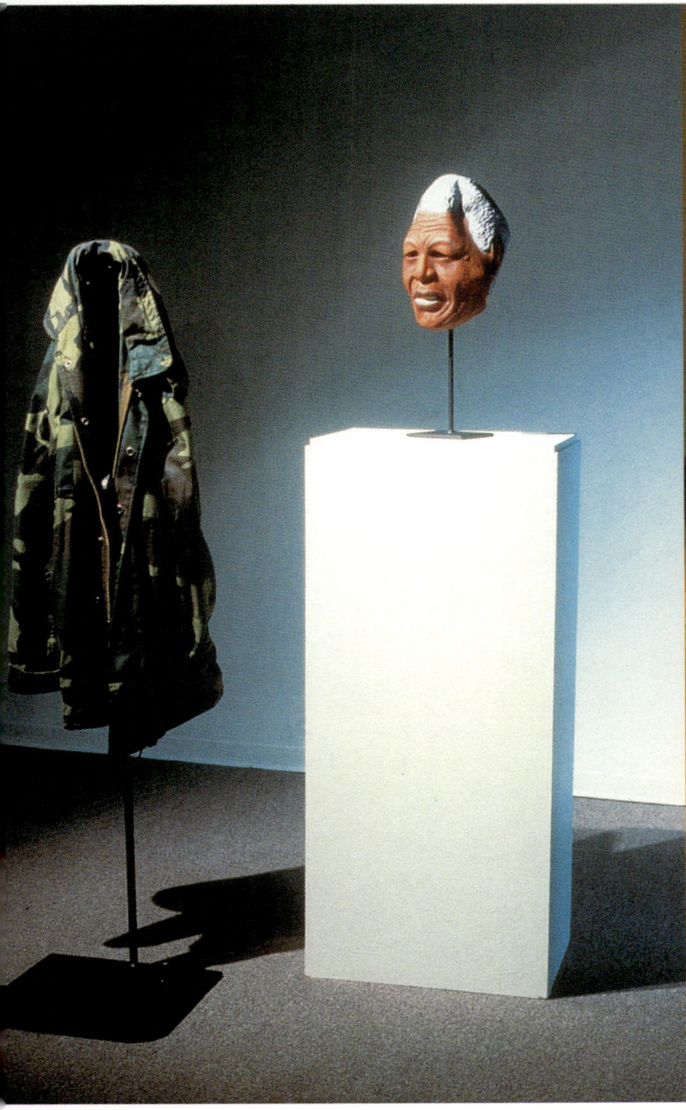

Kendell Geers
*Title Withheld (Out of Africa)*
1997

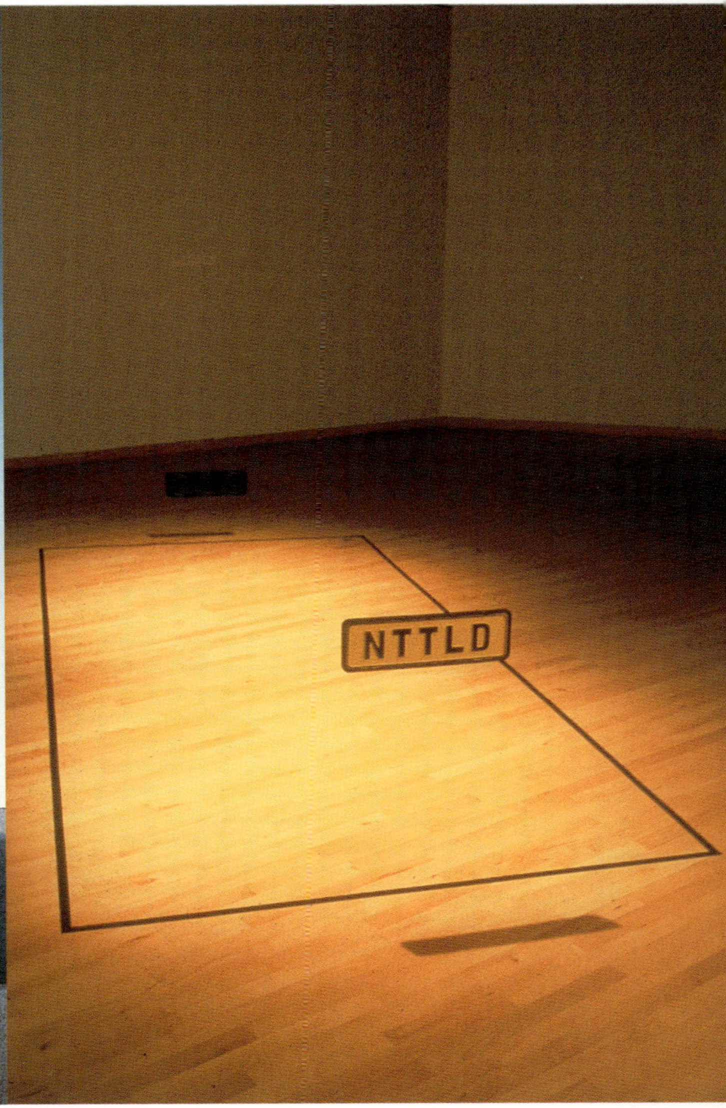


Kendell Geers
*Title Withheld (Untitled)*
1993

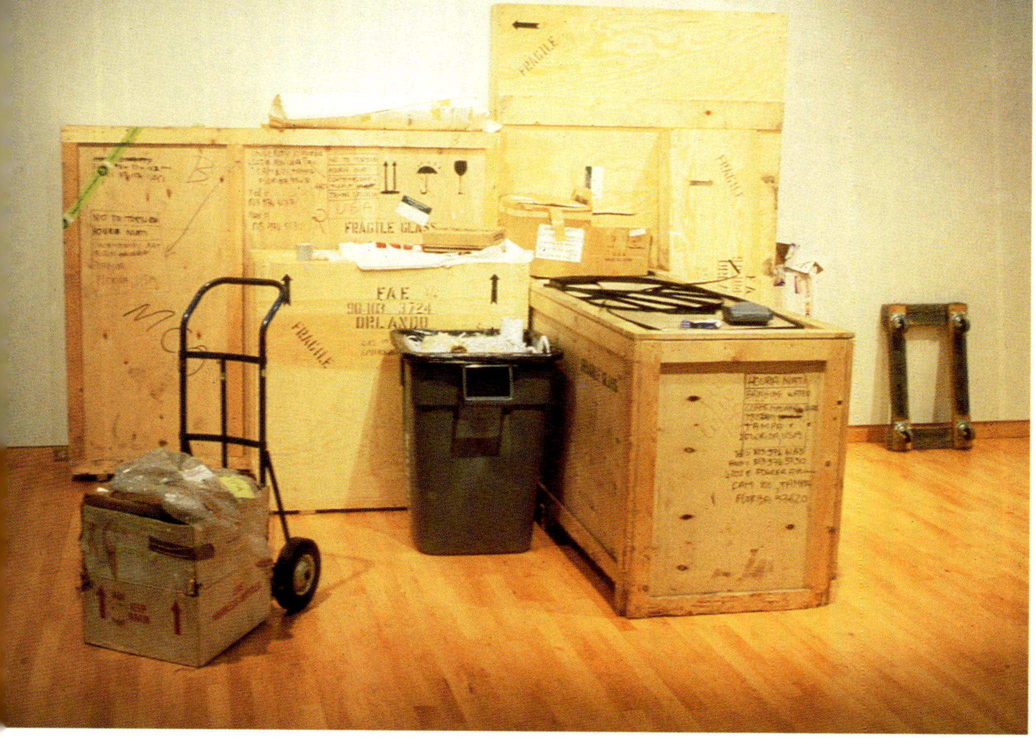


Kendell Geers
*Title Withheld (CROSS/ING)*
1997

Kendell Geers captures the idea of crossing in his work by dragging the object or idea from the space of the ordinary and conventional into that of the rarefied; he pulls his objects across into the arena of the sign. In one of his untitled installations in this exhibition, Geers assembles crates and containers in which the works in the exhibition were transported, containers that bore not only works but through them their makers and the visual cultures and histories they represent. Geers's containers, originally conceptualized for the 5th Bienal of Havana, serve as both surrogates and witness, as facilitators and a silent narrative of translocation and transliteration: accomplices in the registration and violation of boundaries.

In a 1997 installation at the Museum of African American Art, Tampa, Geers posed a mantled rubber mask of the venerable Nelson Mandela and a camouflage jacket on a pedestal, all across a divide from the museum's artifact shop with its stock of African masks. The Mandela mask, as both artifact and irreverent icon, speaks to both the deification of its referent and to the uniqueness of the democracy which that referent and his colleagues have made possible in South Africa, perhaps the only country on the African continent where a comic reference to the head of state is an acceptable feature of a free society. The juxtaposition of the camouflage itself speaks, at least in my interpretation, to the fascinating crossings of history through which a demonized terrorist effects a transition over time and struggle to become an icon. Mandela, the condemned leader of a supposedly violent terrorist campaign, eventually transfigures into a saint, as have Menachem Begin, Yasser Arafat, Fidel Castro, and a few others in our time. In time the image crosses over, and the shunned becomes the adulated.[6]

Geers was born in South Africa, where for an entire century the very act of crossing, of stepping across the lines from one territory to another or one group or class or conviction to the next, was at best a singularly bodacious undertaking fraught with risks and liable to dire consequences; at worst, it was virtually impossible. Under the racist laws that for so long governed the country it was forbidden for people other than Caucasians to move across municipal boundaries without the approval or damnation of the law. For each such person, then, to live was to bear the enormously weighty cross of the territory—and racial category—of their legal designation, and many a man and woman and child gave their lives to escape, or were broken and crushed under the forbidding weight of this cross. On the other side of these boundaries was another, the wall of fear that stood between the privileged and the oppressed and across which the former could only venture at equally great risk, no matter how liberal-minded they were. In his classic *My Traitor's Heart*,[7] Rian Malan, a runaway member of the ruling clan, described the terror of this Golgotha and the folly of venturing to cross. "When the chips were down," wrote the young Malan, "and the killing started, there were no whites on the black side of the barricade. None. Ever." Malan and the many who dared to cross, from either side, were inevitably made to pay.

Geers grew up at a moment in the history of his country when the specter of this cross, though on its wane, was nevertheless present. In time the official policy of racial boundaries in that country has fallen to history, but the habits and sensitivities and retentions and vestiges remain, and to cross is still an act of courage. In his art Kendell Geers has constantly made that journey, sometimes just to prove that it can be made, if only in the name of art. While apartheid was in the throes of death, Geers took the unusual step of registering himself as a member of parties from across the ideological divides of what was still a country of extreme convictions. He joined the liberal coalitions of the liberation struggle as well as the extreme right-wing, white nationalist parties, and the symbols of these conflicting allegiances—his party membership cards—he made into a work of art. Seen at worst as a cynical commentary on political faith, and one that undermined its own goals by relying on the unshared privileges of his race since it could not be undertaken by any other than a white, Geers's act was nevertheless risk-laden, one for which, under certain circumstances, he could pay with his life. It was also a venture into areas of art-making barely validated in the West, where artistic projects of such a political nature were at best marginalized, at worst dismissed. It was anticonvention and politically incorrect, yet it spoke to a strong belief in the rights of the creative individual to step across lines and, even at the risk of controversy, to see through the limitations of narrow historical positions onto the larger stages of the human condition.

Geers's mask at the Museum of African American Art began its life in this terrain of precarious crossing, appearing first as part of a self-portrait of the artist dressed as the soldier Mandela, and proposed for exhibitions in South Africa. Though it generated mild controversy in that form, as well as accusations of profanity, it was nevertheless a valid reminder of the power of time and triumph to gloss over the imperfect perceptions and convictions of the past. It also put the touted freedoms of the post-apartheid state to test, revealing in the process vestiges and residues of old sensitivities that only time may erase. In his other work Geers continues to move back and forth, creating works of extreme subtlety at one moment, engaging in creative acts of violence at another, insisting on the limitlessness of artistic privilege, pushing the boundaries of aesthetic perception, challenging us at every stage to find the courage to cross and in doing so to discover how enriching—though often frightening—it can be.

LUBAINA HIMID

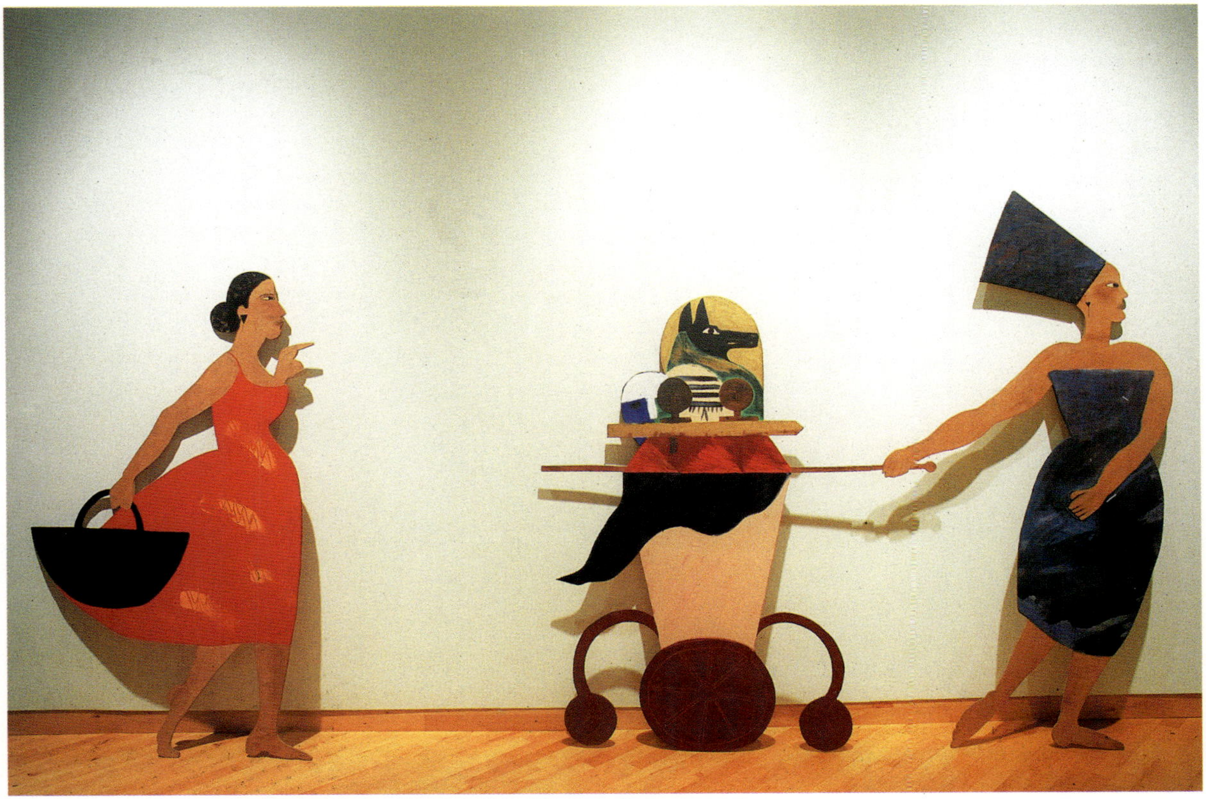

Lubaina Himid
*Restoring the Balance*
1987

Lubaina Himid
*African Garden: Billie Holiday, Mashulan, Augusta Savage*
1993

Lubaina Himid was born in Zanzibar, Tanzania, and was brought up in England, testimony to a life of embarkations and arrivals that she beautifully describes in her essay "Beach House."[8] Since 1983 she has practiced and shown her work as an artist all over the world. In England she played a key role in the 1990s as a cultural activist, helping to call attention to the work of young artists of African descent in Europe and to the questions and issues that not only preoccupied them but had relevance to society at large. Himid's *African Garden* (1993) refers to the cultivation of memory and the restitution of history through a reinsertive intervention whereby figures and moments of significance are restored to their place in our commemorative narratives. Billie Holiday, Mashulan, Augusta Savage—icons of the black cultural heritage whose places in history have been all but erased by mainstream revisionism—take their place again in the garden of memory. In *Restoring the Balance* (1987), Himid traces the history of the black female from ancient Egypt to the present, in this chronology also tracing her own story as an African woman whose sense of self anchors in a knowledge of the history and place of her predecessors. From Hatsheput to Himid a long line is imaginatively drawn that reminds us of Margaret Busby's *Daughters of Africa*, a long line of women who write indelibly of the resilience and richness of a continent and people.

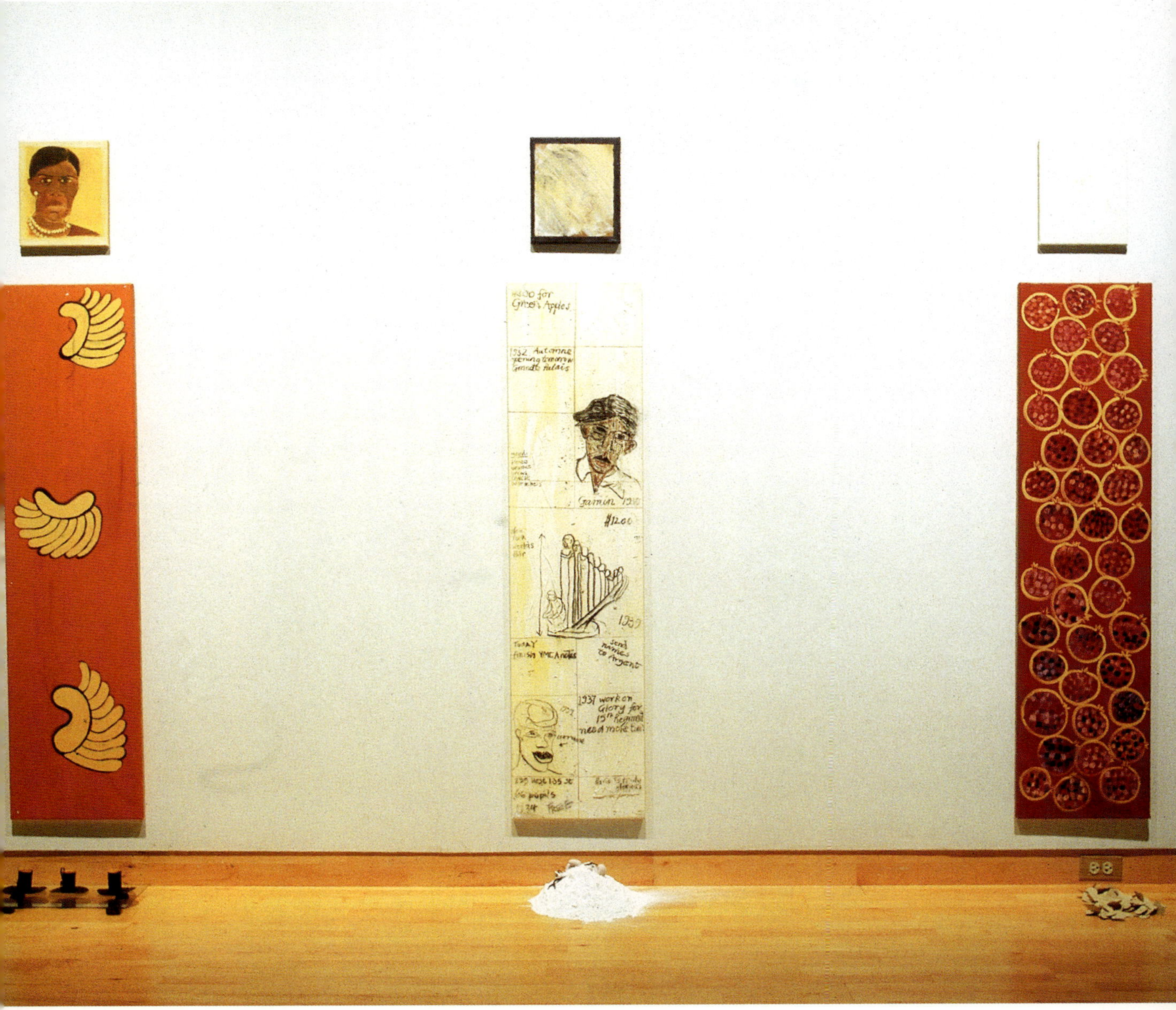

Gamin 1930
#1200
1939
TODAY finish YMCA notes
send names to Argent
1937 work on Glory for 15th Regiment need more time

MARCIA KURE

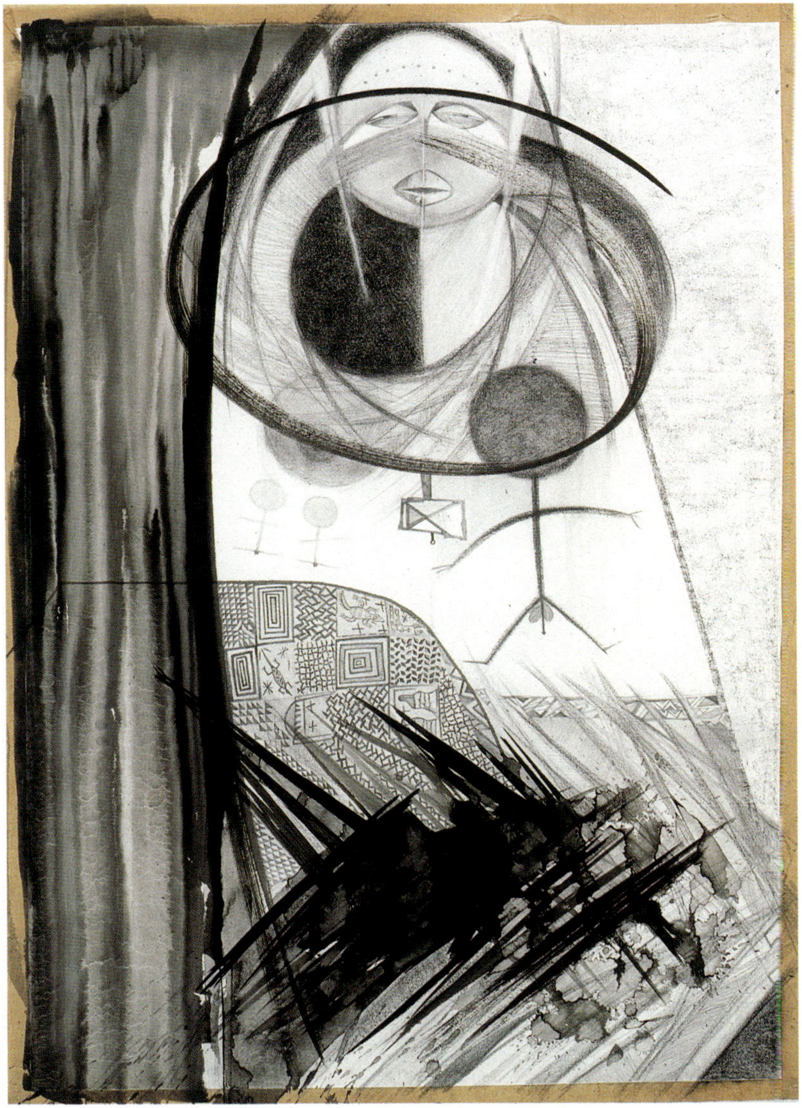

Marcia Kure
*Woman Suckling Her Male Child*
1996

Born in Nigeria, where she received her education at the highly reputed fine and applied arts department of the University of Nigeria, Nsukka, Marcia Kure has already distinguished her practice through a determined devotion to the centering of women in her work. In her paintings and drawings not only are women the protagonists, they are portrayed with a sensitivity and attentiveness to the details of their natural and social conditions that is rare in the work of her contemporaries.

A fine example of Kure's admirable depth is her "Woman" series, especially *Woman Suckling Her Male Child* (1996), a particularly loaded work in which the artist addresses the interesting social and psychological implications of maternity and the act of breast-feeding. In *Woman Suckling Her Male Child* there is a nonprecise, almost ambivalent depiction of the protagonist's experience in the act—a countenance deliberately sited between pain and unpronounceable ecstasy, between the burden of a chore and the unabashed thrill of a secret, disallowed indulgence. In part we are reminded of the character Ruth in Toni Morrison's *Song of Solomon,* a deprived and frustrated wife who in order to compensate for her rejection at the hands of her hard and insensitive husband relishes the experience of suckling her slightly overage son. Morrison describes Ruth's practice as "one of her two secret indulgences"[9] and indicates that Ruth derives her pleasure not from mere maternal joy but from the thought of how grown her son is, from knowledge of the element of taboo in her indulgence. Hers is the thrill of the forbidden. As we look at Ruth, we are reminded of how the fine line between a perfunctory maternal act and its sexual corollary is constituted almost entirely by age and distance. We are reminded, also, how nurture and conjugation overlap both physically and metaphorically; in this instance the oft-unacknowledged or repressed eroticism of the act is restored and heightened by the cross-gender nature of Ruth's secret preoccupation. Eventually her son acquires the rather derogatory name of Milkman, a tag that registers society's ambivalence toward the act.

Kure's *Woman Suckling Her Male Child* specifically calls our attention to the sexual implications of suckling as an erotic act by its deliberate choice of a cross-gender experience, without necessarily eliding the possibility and implications of the erotic in maternal relations within the gender. And just like Morrison, Kure makes allowances within her depiction for an understanding of suckling as an affirming act, as an important territory of privacy, as both a condition and a privilege. Her election of such an uncommon theme is both unique in the context of Nigerian women's art and bodacious, for it is a departure from the bland preoccupation with the stereotypical and uncritical which dominates the practice of her contemporaries, just as it introduces an important theme into the anthology of contemporary feminist art practice.

Kure addresses other questions that relate to the narratives and location of women in her society. In *Woman in Purdah* (1997) and *Woman at a Wedding* (1996), for instance, she touches upon sensitive issues of the dialectics of culture and regulation through social expectation, with as much ambivalence as leading silence. In *Woman as Warrior* (1997) she presents a female figure whose shield confronts us with thought-provoking, deeply encoded physiological allusions, as well as less-encrypted historical references: Kure sets out to recover the history of powerful African women, figures such as Amina of Zaria, Tinubu of Lagos, Mandela of South Africa, or Sheba of Abyssinia, women who rode into battle at the head of armies or led popular struggles. This project of recovery and restitution is one of the artist's dear preoccupations. Beyond the idea of power, Kure's metaphor of the warrior also reminds us of the fact that for the woman in a world that has only grudgingly begun to concede equal humanity across the genders, to live from one moment to the next is a battle.

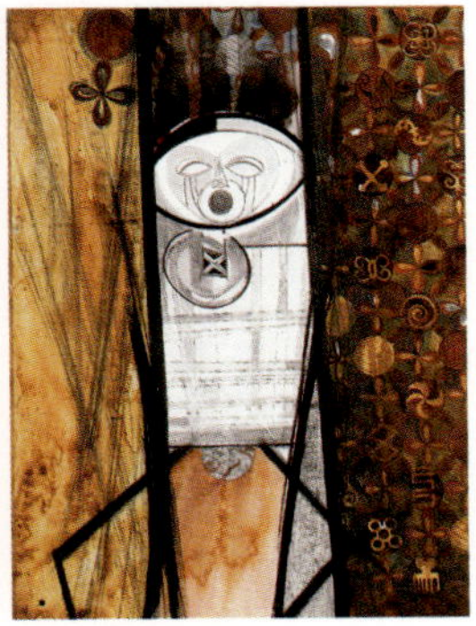

Marcia Kure
*Woman Giving Birth*
1997

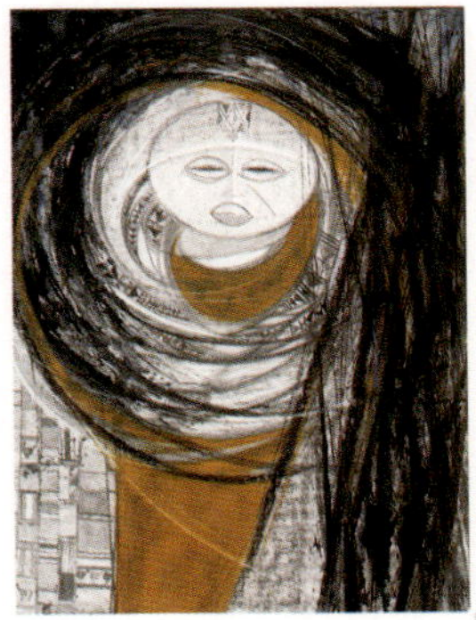

Marcia Kure
*Woman as Akuaba*
1997

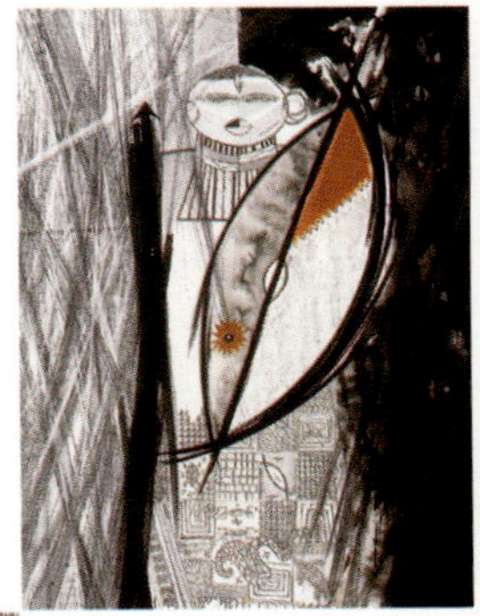

Marcia Kure
*Woman as Warrior*
1997

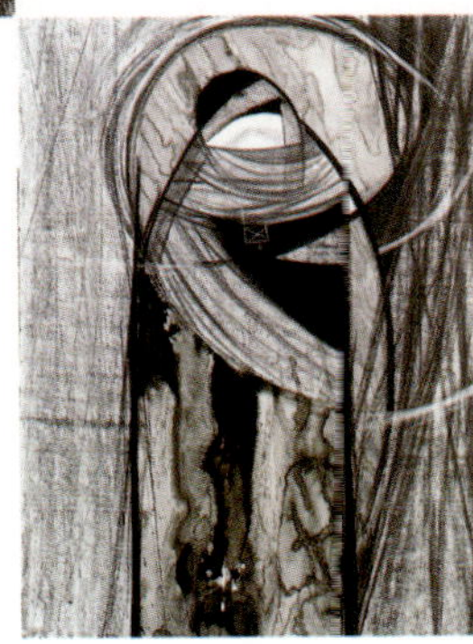

Marcia Kure
*Woman in Purdah II*
1997

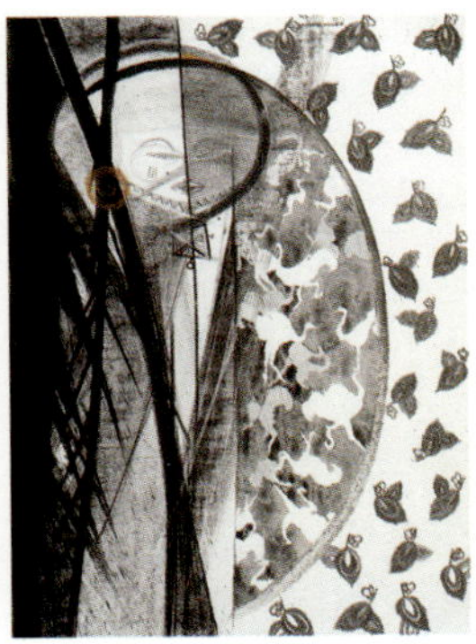

Marcia Kure
*Woman at a Wedding*
1996

(all details)

HOURIA NIATI

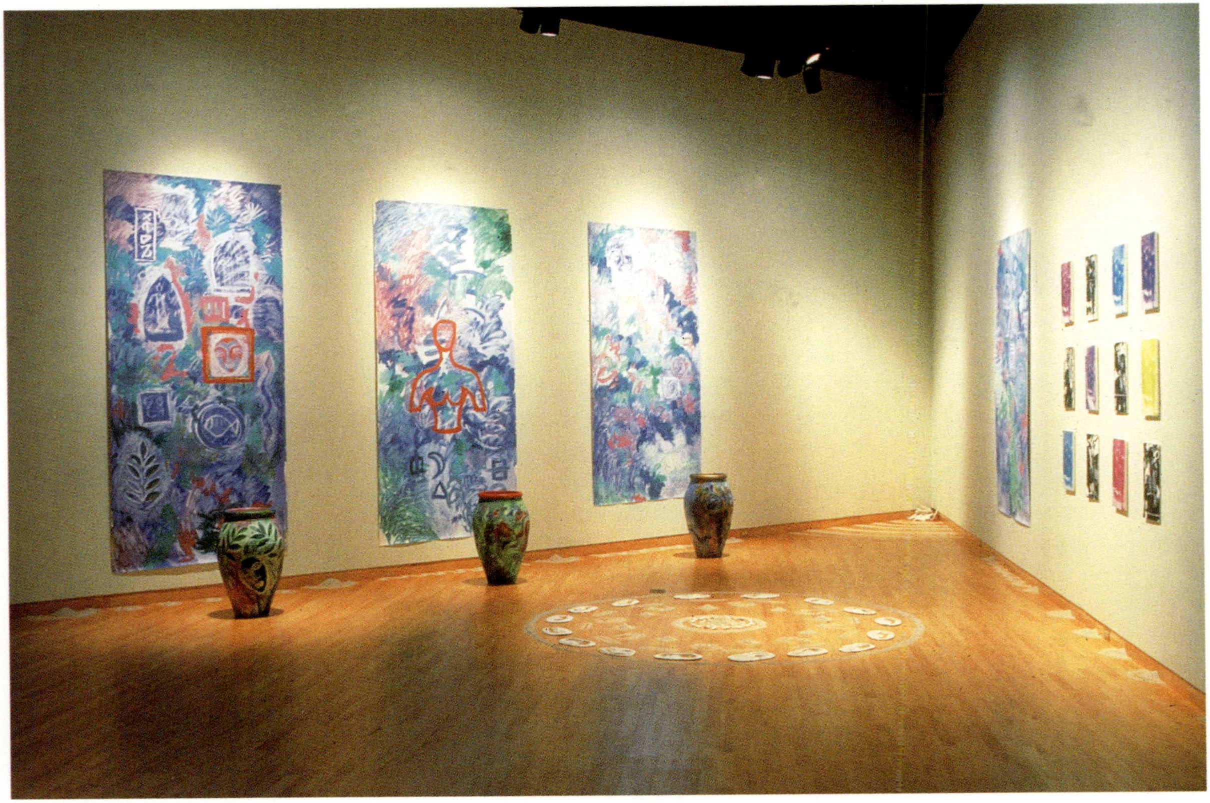

Houria Niati
*Bringing Water from the Fountain Has Nothing Romantic about It*
1991

## BRINGING WATER FROM THE FOUNTAIN HAS NOTHING ROMANTIC ABOUT IT

Houria Niati

I begin a story which could
never exist
In the obscure meanderings
Returning with fantastic
Repetition to some civilizations
The strength of pages
Which turn with a multicolored
rhythm
Of the prodigious
Inventions
The stories are told but
They do not explode
Because they fuse
Silently to the oppressed spirits
And are engulfed by a cavernous
defeat
Which loses itself and is forgotten
Why do the travelers look for
total accomplishment
of the absolutely unreal or
Provisional separation
Bursting in the veins
Like nonfatal poison
Evidence is not truth
Because the vision can change
So why idolize
The bodies without feeling
Which swollen by unconscious
dissatisfaction refuse to rebel
And accept the invigorating
Neutrality of the objects
Too much wanted and
misunderstood
Traps opening
Like enormous mouths
But the pleasures are transient
The immobility is latent
Aggression and trembling
Agony
In the night obscured by
The Ignorance and great
Beauty of passing sirens
Belonging to the hereditary
Power
Of the Man who holds the keys
Of destiny
The immobility is the strength
Fragile and compassionate
Of the Mothers columns of the
descendents
Messengers from the womb
Of the obligatory reproductions
The immobility is the serenity
Contagious the silence in a
World where the explosion of
Revolution blocked by
The walls built by
Possessive hands
The immobility is the repressed dream
Of the impossible escape
To far horizons
That disappear into thickness
Of the endless night
Not thinking is to accept
The conditions with no conditions
Not thinking is to burst out laughing
Like a bomb
Obscured by the night
By the incredible misadventure
Of limited freedom
No matter what the silence
In the illuminated darkness
Accepting the pleasures of the dark of
night
And rejecting the immaculate morning
Who are you women of the mountains
Nymphs of allures who
Amaze the hardest hearts
Do you really exist
Have you really felt
The eyes of this Artist on your bodies
Who bravely
Wanted to capture
This ecstasy
Not only the
Repression and Exploitation
Who are you Women who submit
To sensual passion
In the shadowy houses
With half opened windows
Looking into interior courtyards
Women fatal and mysterious
Powerful in their innocence
Out of the ordinary out of time
But who shut their eyes to
The magical thread of
Racial multiplicity
They are the dream
of a fragile mystery
Which has traveled
Desperately
Searching for peace
And newness
Deformed Orientalist
In the beginning of a destruction
Inevitable but positive

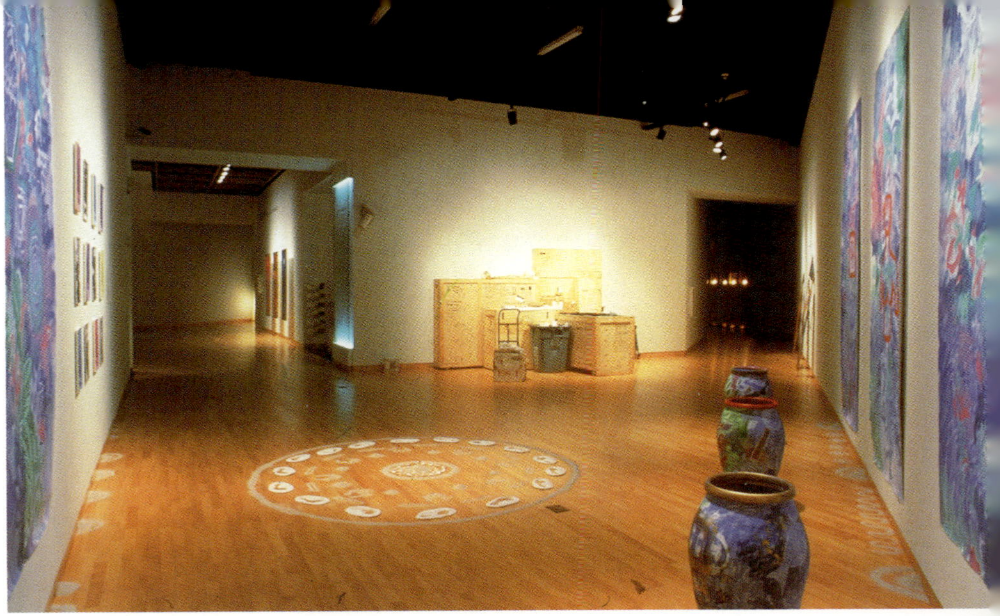

*foreground:* Houria Niati
*Bringing Water from the Fountain Has Nothing Romantic about It*
east wing
1991

*background:* Kendell Geers
*Title Withheld (CROSS/ING)*
1997

Houria Niati, an Algerian who has lived and worked in Europe since the 1970s, is particularly significant among contemporary African women artists for her combination of a personal history of political involvement, an impressive knowledge of the history and politics of representation, and a practice dedicated to the critique and negation of male and colonial traditions of female imagery. Added to these, her importance also derives from her identification of the correlations between colonial desire and power, and male hegemony within the space of image-making, and how both came together in colonial imagery to produce some of the most problematic depictions of women. In her paintings and installations Niati goes beyond the deracialized frames of orthodox feminist art and critical practice to reveal issues and spaces of complicity as yet unattended: those spaces where the African or Arab woman is the objectified other of not only the male but also of the European, spaces of racial voyeurism and colonial license where gender is as relevant as geography.

In some of her paintings and installations Niati takes on nineteenth-century Orientalist painting and photography in which the women of North Africa are encapsulated in a romanticized sensuality and langour that erases their subjectivity and occludes their realities. In *No to Torture* (1982–83), a reinterpretation of Eugène Delacroix's *Women of Algiers,* Niati deconstructs Delacroix's false and voyeuristic representation of Arab women by carefully analyzing the iconography of Delacroix's paintings to produce what for a better term we might call an x-ray image of the original. Niati replaces the smooth finish and veneer of Delacroix's painting with her rough and coarse brushwork, and in place of the French painter's carefully manipulated chiaroscuro she produces a stark contrast in polychrome. She cancels or rubs out the faces of the women in her interpretation to indicate how in the French painting the selfhood of the characters is erased and the focus is placed on their bodies, how in Delacroix's *Women* the viewer's attention is drawn through composition and the use of light to the bodies of the women—to their bare feet and exposed bosoms, to their stares encrusted with sexual longing and invitation, to elements and features that effectively transform them into objects of erotic fantasy.

Where Delacroix located his figures in the dim and luscious architectural ambience of Arab opulence, the luxurious and dark comfort of the harem, Niati restores them to the stark particularities of reality against which they appear skeletal and stripped. Scholars have shown

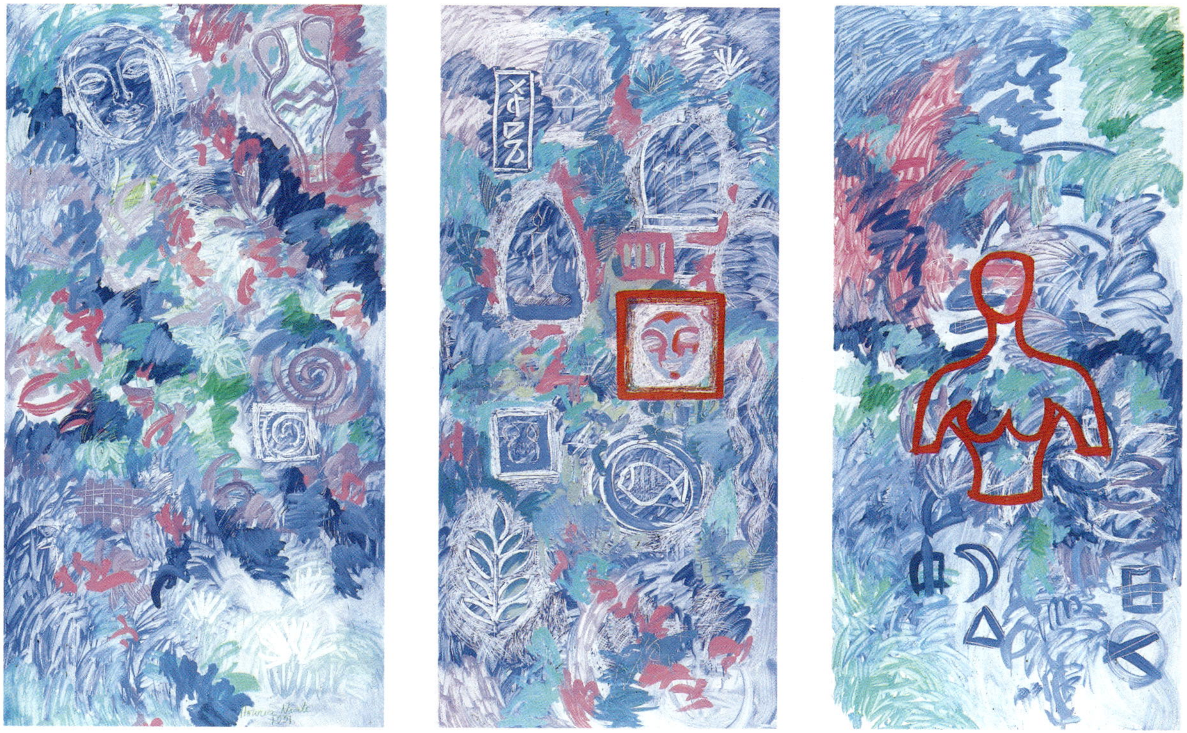

Hour a Niati
*Bringing Water from the Fountain*
*Has Nothing Romantic about It* (detail)
1991

that Delacroix's representations of North Africa derived more from orthodox fictions of Orientalist imagery than from the realities he observed in his North African travels as contained in his notebooks.[10] With little knowledge of life in the harem or indeed the North African family, European artists constructed images born out of a mixture of projected desires and fantasies and a will to otherize through denigration.[11] Constructed as nothing but a sexual entity, the Arab woman in colonial representation is not only an object of desire but representative of a less-civilized culture marked by the prevalence of unrestrained sensuality and baseness. The women's bare feet, both erotically charged and culturally codified with associations of excess and abjection—bare feet being a signifier of the wild sexuality of the lower classes in French society at the time—mark them and their race as occupants of a lower rung in the hierarchy of races upon which the colonial will to power was itself founded.

What popular desires painting was unable to fulfill due to its elitist nature, photography certainly attended to, spawning an entire industry in colonial images, some of which bordered on pornography. With photography (and particularly the postcard), the fabled sight of Arab women's flesh was brought within reach of the ordinary European male, and the more tantalizing the images that made this possible were, the more desirable they proved and the bigger the industry grew. In all of this, as in painting, the actual women of North Africa were not in the least considered; the sole focus of the distant gaze was on their marketed and consumed bodies. One popular genre of this industry that Houria Niati has confronted is the happy Arab maiden with a water pitcher featured on French postcards throughout the late nineteenth century. The genre was characterized not only by the false image of the happy water-maid, but also by a methodical pornographic intent that often showed these women naked with their breasts exposed and carefully accentuated by the curves of their pitcher, which in turn they were made to position in a manner that directed the eyes along its length to their waist and crotch. In less explicit examples the women were nevertheless shown with a demeanor that belied the strain and discomfort of water-bearing. In effect, not only were they taken advantage of through the forceful trading of their bodies made possible by colonial license, but their true plights were equally erased in order to satisfy the requirements of the romantic colonial imagination.

Niati challenges this double infliction of both annexation and erasure in her installation *Fetching Water from the Well Has Nothing Romantic about It*. In the installation she brings together large-scale drawings in which she repeats her strategy of posing roughness against colonial gloss; enlargements of the colonial postcards discussed above; and real water pitchers that make it possible for the viewer to revisit the image of the happy-go-lucky water-bearer and question the possibility of experiencing glee under the weight of those pitchers. On the pitchers Niati also draws images of the women with an anatomical ordering that reminds the viewer that the colonial postcard image of the North African woman had little to do with the condition of the woman and had everything to do with desire for her body. Although Niati's pitchers are decorated and colorfully transformed, what we are made to appreciate and contend with is their sheer weight, their unbearability, and by extension the unbearability of the life and social locale for which they are signifiers. The artist makes us deal with the facts of a society in which women are made to bear that weight. She forces us to share, symbolically and psychically, in that weight, as in the moral weight of a colonial project that romanticized and aestheticized it. Niati's women of Algiers are not the smiling water-bearers and sexual objects that we are introduced to by outsider imagery, nor do they reside in the languid heat of sexual availability and frustration from which the European male may rescue them; the lives they lead can be harsh and demanding and filled with hardship and drudgery, just as are the lives of women elsewhere, and it is this reality, this truth, that Niati forces us to acknowledge. It is these women's lives and vulnerabilities to history that are at the center of her art.

OLU OGUIBE

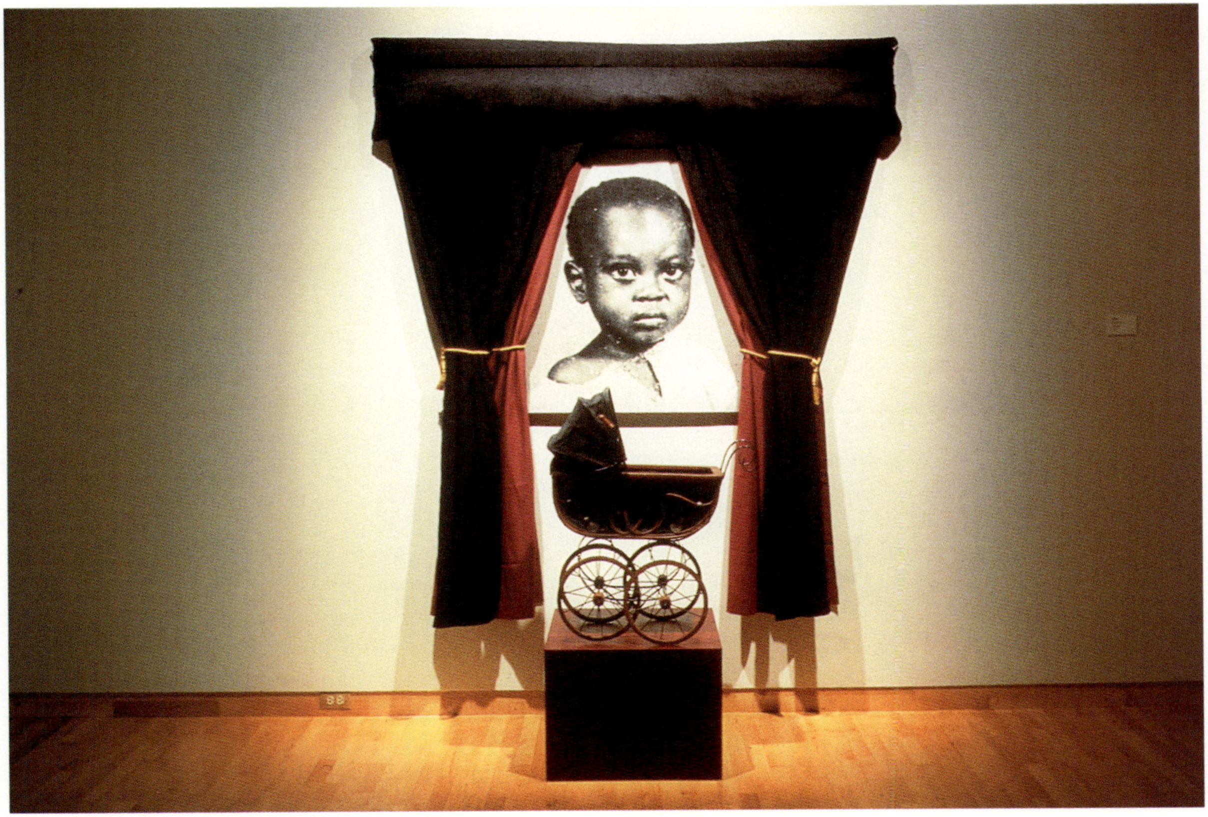

Olu Oguibe
*Buggy (For an Unknown Child)*
1997

Born into political turmoil and raised in the midst of what many consider the most devastating war in Africa's history, the Nigeria-Biafra war of 1967–1970, mine has always been a life of constant shifts between stations and roles, nationalities and allegiances, expectations and convictions, traditions and modernities. Not the least significant of these demanding shifts is my location within the often awkward but nevertheless long tradition of the artist-critic, traceable at least to that maverick and often diabolical critic, the English painter William Hogarth, and in addition, my alternations between creative writing and theory and between poetry and visual art, and, indeed, the fact of my descent from a long line of deity priests. This lineage ended with my father's defection to the Christian ministry, which exposed me at a very young age to the entire history of the Church, and led to my progression from being a Sunday School teacher in my early teens to a stint as a lay-preacher in my mid- and late teens, to loss of faith at age twenty-one.[12]

For a long while my work as an artist was preoccupied with a fastidious formal inquiry that at one point led to the disavowal of all media and forms associated with the Western tradition, my artistic engagement having a direct relationship with my active involvement in social causes and national politics. For a while, also, the social focus of my work leaned on an exploration of broad thematic and abstract ideas, almost in the romanticist tradition, even when these were hinged on specific historical incidents or circumstances. Beyond these, however, a careful survey of my oeuvre over the years reveals a more enduring, perhaps more endearing, preoccupation, namely the recurrence of personal narratives of trauma, and a quest for resolution and closure through commemoration. It is this that defines my production even more firmly in the mid- and late nineteen-nineties.

In one of the earliest pieces in the commemorative genre, *Passage (Through a Child's Eye)* (1985), a load-bearing train of fleeing war refugees courses through the bewildered head of a child, his wide eyes and taut countenance registering most clearly the horror of witness and the trauma of recollection. In *Passage* I make direct reference to my experience as a war-child, an experience to which I return in the essay "Tragedy and the Artist: Remembering Biafra." The trauma of growing up in a war that by some accounts claimed more than two million lives, a war in which half of my generation was destroyed either through a consciously engineered famine or through indiscriminate military violence, left me with five straight years of daily nightmares as well as numerous other uncertain terrors, all of which in time have coalesced in a rhetoric of exorcism through reiterative invocation of the horror and loneliness of tragedy and loss. In the first of his *Five Lectures on Psycholoanalysis* at Clark University in 1909, Freud referred to the recollection of the traumatic as *reminiscences,* and to those images, objects, and moments that commemorate them as *mnemonic symbols,* props and signs in which the kernel of tragedy is memorialized and thus put to rest.[13] Through such monuments and symbols the memory of trauma is externalized and exorcised, saving both individual and society from the path to neurosis. Even more importantly, they serve to remind us that should we find the diligence to heed their appeal and gestures, we remember the terror of the past and of the necessity of a present and future in which that past is not repeated.

In a series of paintings from 1989 I began an outward shift toward a more universal exploration of the meaning and nature of loss and commemoration, outside of my own personal experience, indeed outside the grand context and historical specificity of war. In *The Bereaved* (1989), *Orphan* (1989), and *He was the only son and his mother was a widow* (1989), my goal was to visualize—and in so doing seek to understand—the experience of intimate loss at the more important level of the immediate family, the direct bearers of the burden of loss rather than the removed theaters of public monuments and national mourning. This exploration would lead eventually to the 1995 installation *Oklahoma,* commissioned and

Olu Oguibe
*Memorial to a Youth*
1997

produced in the wake of the bombing incident that took place in the city of its title. In a world in which violence and cataclysms and the terrors of war claim hundreds of lives daily, or indeed, in a country whose history is replete with violence and genocide, the incident in Oklahoma was not, of course, a particularly unique one. The installation, however, was an opportunity to bring home to a society grossly desensitized through the glamourization of violence the reality of pain at the most intimate locales of tragedy and loss.

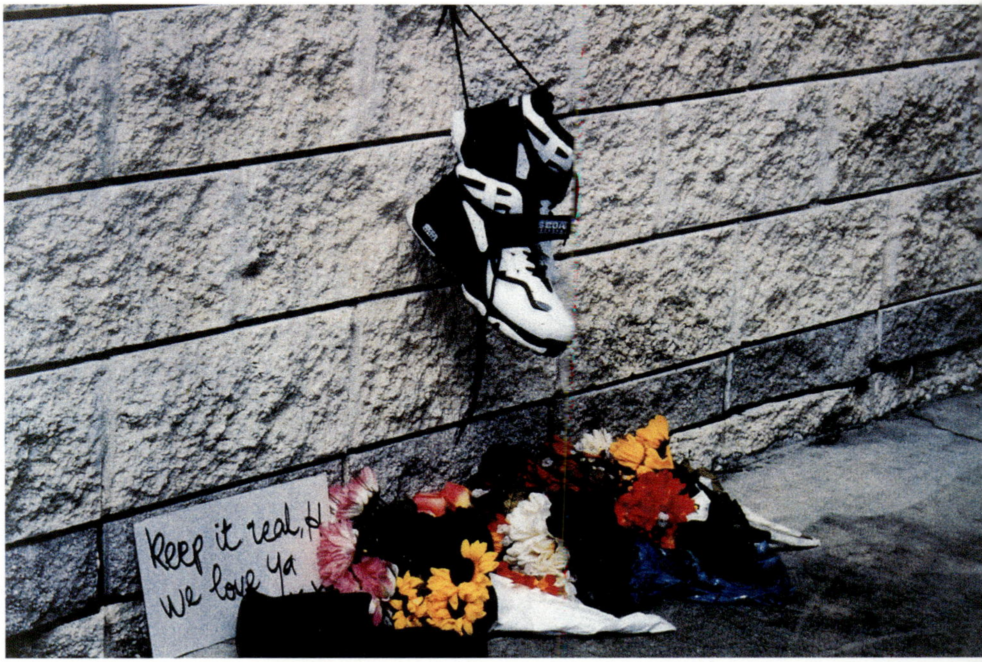


*Buggy (For an Unknown Child)* (1997), *Memorial to a Youth* (1997), and *Mementos* (1996) retract from the public space and the specific historical moment into that of the generic, thus redirecting our focus to the deeper, private meaning of loss. Consisting of a pram or buggy on a pedestal and the enlarged photo-image of a child whose identity is not revealed, *Buggy* reinhabits a portion of the exhibition space with an anonymous presence whose innocent gaze demands reflection and silent contemplation. The pram is antique, and the aim of this is to introduce an element of relevance without temporality; it is a sign of timelessness. The photograph is framed with red and black drapery to create an ambience of somber dignity. Like *Oklahoma*, *Buggy* envelops us with the reverentiality of a hallowed space, pulling us out of the nonchalance of our everyday existence and ensuring that we concede a moment of silence in its presence. In this moment of reverence we are able to understand grief anew, to regain our ability to empathize with the afflicted, to dig inside ourselves and reach that core of our humanity where we are able, once again, to feel for a stranger; to know mourning as a communal experience, to rejoice with those who rejoice and grieve with those who are afflicted. The child in *Buggy* is every mother's child, the experience of lingering presence and the affirming transition from grief to celebration is one we are able to share, one we may partake of, because despite the alienation and segregation and desensitization that have today become prevailing parameters of our lives, there is a corner deep inside us all where we are still able to connect and share in each other's moments of rapture and distress.

TRACEY ROSE

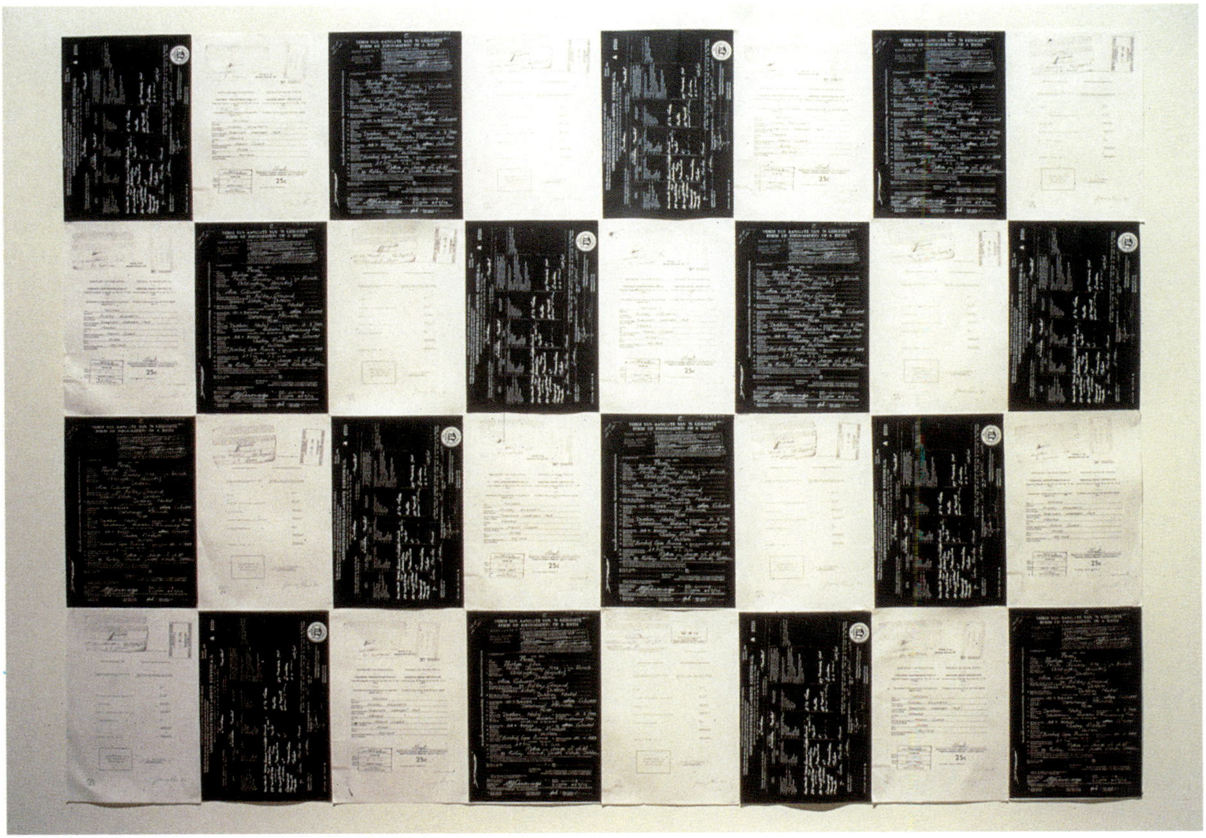

Tracey Rose
*Untitled 1940–1996*
1997

Tracey Rose, the youngest artist in this exhibition, was born at the very height of the terrors of the apartheid policy in South Africa, only two years away from one of the great political and human tragedies of this century, the Soweto uprising of 1976. Rose was born "colored," a nebulous designation in apartheid's shady categories of race and belonging that fell between "white" and "African." To fit onto this often indeterminate rung of the ladder of hierarchies was not only to know substantially fewer privileges and freedoms than whites and considerably more than blacks, but also to be caught in between two particularly antagonistic designations and territories that met at the median of distrust and hate. At this middle point, every nuance of physiological peculiarity counted towards one's secure sense of belonging and one's place in the matrix of things. In an untitled video installation (1997) Rose shaves off her body hair to the deliberately irritating drone of a hair clipper, taking us into this slippery terrain of race and politics where the texture of hair and the hue of skin are crucial determinants of an individual's place and fate.

Rose reminds us of the potent symbolism of the one feature of our physique which almost always commands our greatest attention, and in the case of racial societies, of society as a whole. Throughout history the significance of hair as a social, even political sign has been reiterated: groups and individuals have been either singled out because of their hair and subjected to persecution, or they have used their hair as a theater of struggle and self-affirmation. In our own century we are reminded of fascist Germany where the texture of an individual's hair could condemn them to the horror of the gas chambers, and where, just as horrifyingly, the hair of the persecuted was in turn made into a commodity to be recycled and traded. We hear Paul Celan in that most harrowing of reminders, the "Deathfugue," in which the poet, himself a survivor of the Holocaust, encapsulates the horror and tragedy of the Jewish experience through the imagery of hair when he writes: "your golden hair, Margarete/ Your ashen hair, Shulamith." Margarete the Aryan princess keeps her golden hair while Shulamith the Jewess, incinerated for her race, loses hers in a heap of ash.[14]

We are reminded also of the other Holocaust, that of the African-American experience during which the texture of one's hair was enough to place one within or beyond the reach of social molestation and penury. In time African-Americans made hair into one of the most potent political signs of the twentieth century when, at the height of the civil rights movement, they turned their hair into a tool of affirmation. The long-grown Afro became both a cultural and political statement serving not only as a relink to Africa through the affirmation of Negroness, but also a negation of a history of deprecation of that identity. As hair was thus repoliticized, other affirmatory movements would elect to use it as a sign: the pacifist, flower-child generation in America; the early feminist movement with its shorn hair; and in time, the hip-hop generation of African-American men for whom baldness signifies, in the same manner as grown hair did for the Black Power generation, a reclamation of their rights over their bodies and destinies.

In Rose's video installation she meticulously shears her body hair before us, a gesture which, as implied above, speaks to a particularly potent social posture. Shorn of our body hair, whether straight or curly, blond or dark, flowing or woolly, we all are returned to the basic physiognomy that defines all of our species, and thus are taken across and beyond the configurations and confines of inhabiting laws based, as Frantz Fanon would say, on the *epidermalization* of difference. Yet the loss of our hair is a veritable loss indeed, and its removal is a most significant ritual of surrender, of shedding, of sacrifice in readiness for transformation—in readiness, as it were, for crossing. When hair becomes a cross that we bear, with its surrender both weight and passion are lost, returning to us the clarity of civility and the incredible lightness of being. In the South Africa in which Rose had her childhood, and perhaps even in the society that has survived it, to lose your hair is to surrender a sign, to sacrifice the fiction of a precise identity,

Tracey Rose
*Untitled*
1997

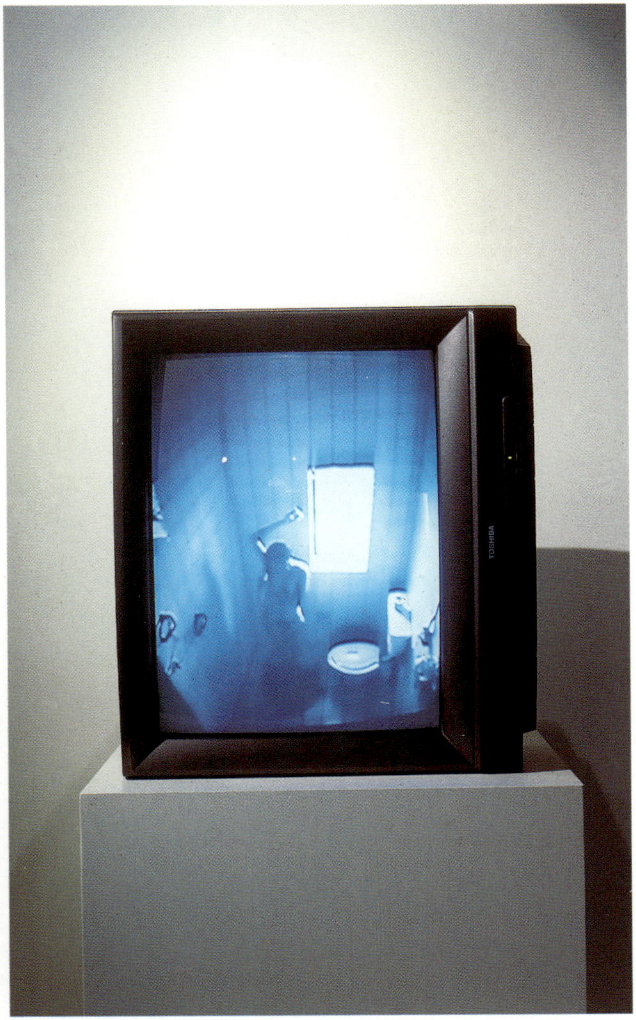

to magnify the incongruence of a particular prejudice upon which, for so long and for so many, life and death so precariously hinged. It becomes a disavowal, a rejection of definitive ciphers, a reclamation of that territory of the undecidable within which life's essence may be found.

In another untitled installation (1997), Rose displays her birth certificates, a reference to the peculiarities of a formerly racial society in which the precise determination of identity held the fate of the individual. From the documents, strategically displayed in a checkerboard pattern of black and white, we learn of Rose's own location in the in-between, her transfigurations and translocations, her resistance to the demand for identity and her will to violate the inviolable and invalidate the definitive. Resistance and struggle have brought her society to this location where aspects of identity and precise definition are subordinated to the basic humanity of all.

FOLAKE SHOGA

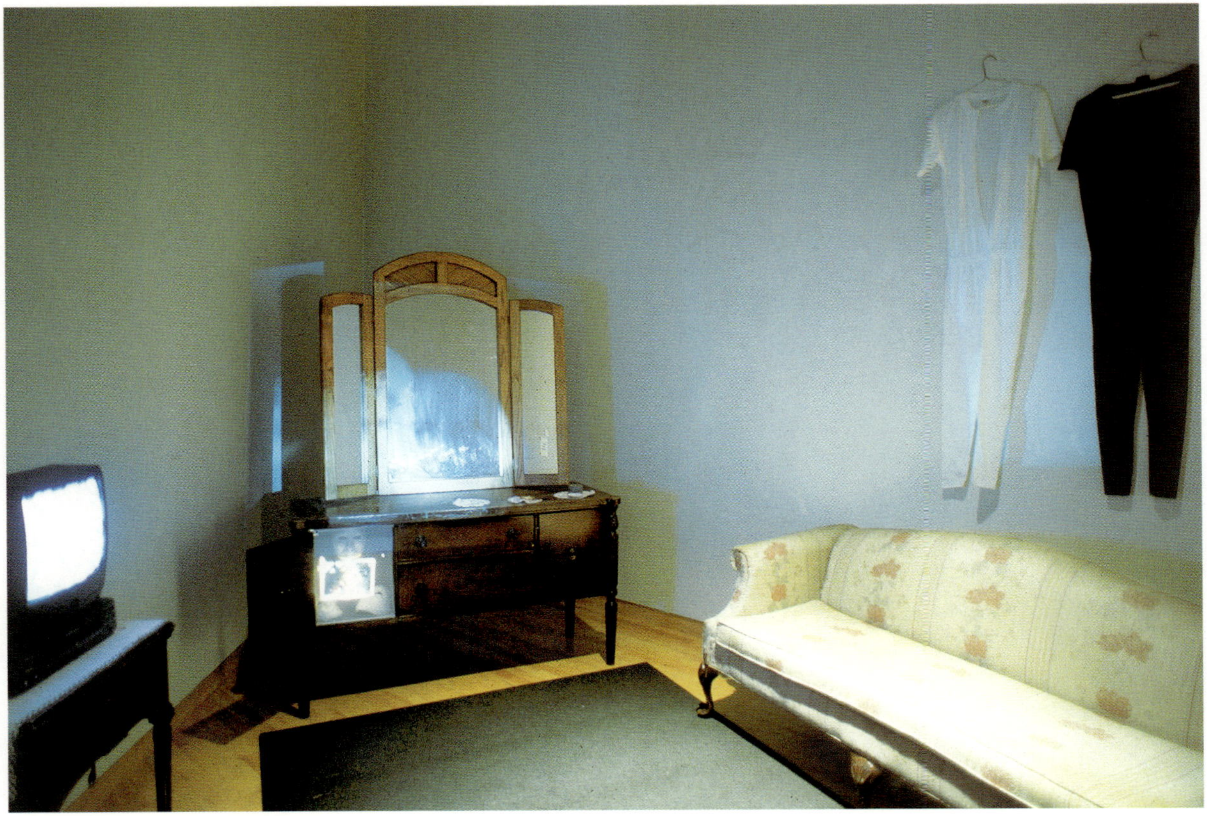

Folake Shoga
*Like Something Not Very Concrete*
1994

Folake Shoga was born in Ibadan, Nigeria, and raised in Britain, where she attended art school and has practiced and won recognition as a maker of short films, an installation artist, and a book illustrator. Shoga deals with the same issues of uncertain identities as the artists above, but she extends our frame to include the contemporary experience of Africans in diaspora. In her elaborate installation *Like Something Not Very Concrete* (1994), she recreates an old working-class living room with winter longjohns hanging on the wall, and cheap but grand furnishings that border on kitsch. There is a television set, that inevitable companion of the lonely and forsaken who live their lives on the short and narrow lane of the dole queue. There is also a large mirror, our clearest metaphor for reflection, and there is a display cabinet inside of which, instead of material treasures and signs of acquisitive contentment—expensive china or rare volumes on philosophy—we find face masks. On the display cabinet the artist projects a series of slides in which different female characters appear, engaged in that most intimate of social preoccupations: hairstyling. We are reminded of Pepón Osorio's *No Crying in the Barbershop,* not for any analogies of theme, but for the peculiar nature of the hairstyling space and moment as a ritual site of vulnerability and confession.

On the huge dressing mirror that frames the living room, Shoga projects images of another ritual in progress: that of masking. As if looking in the mirror to ascertain which countenance fits best, which identity truly belongs to them, as if to discover which role is safest and which most ludicrous, a lineup of women try different masks on their faces and ultimately discard them. It is a ritual of exploration and self-doubt, a journey of self-discovery much like the Stations of the Cross. At every station each character transfigures under a peculiar pressure, a specific expectation, a manner and moment of social intervention, and finding the role into which she is forced undue and unfair to her innermost nature, she sheds that role and its inscribing codes. Except, of course, that which is rejected and discarded is quickly replaced as one mask takes the

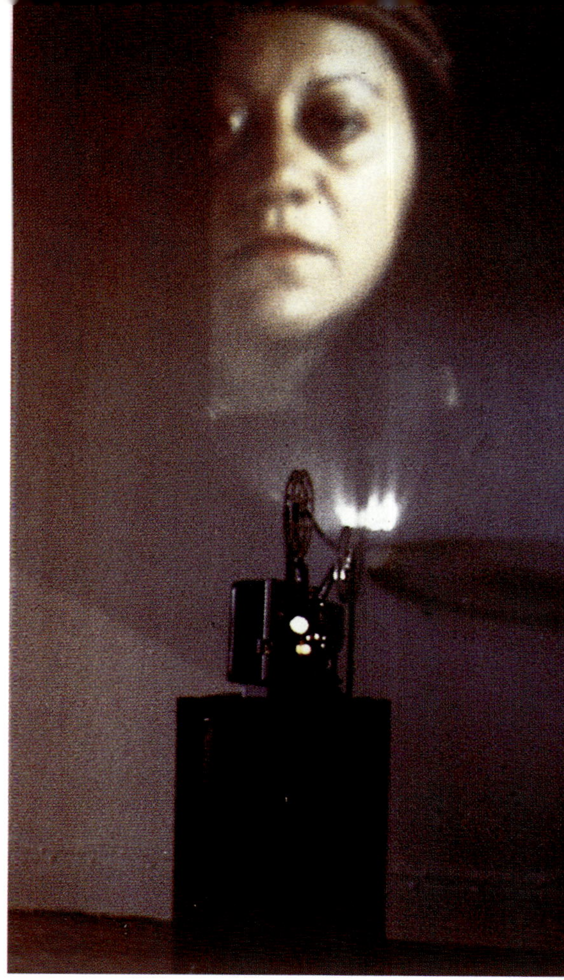

Folake Shoga
*Like Something Not Very Concrete* (detail)
1994

place of another. "People of African descent wear masks," museum director Kinshasa Conwill reminds us in the British documentary *Jean-Michel Basquiat: Shooting Star.*[15] "They wear masks to protect themselves, to assume whatever role society demands of them." But the mask of racial role-playing is also a dangerous and inimical one, for behind this mask lies a fiction of rootedness counteracted and constantly undermined by an insistent anxiety. The social mask, like the racial dialectic that calls it into play, sets out to invest a momentary certainty on a territory otherwise characterized by an inherent and unmitigable ambiguity. It dares to say "I am" when in truth the question lingers: "Who, exactly, am I?" It struggles to invest concreteness on something which by its very nature *is not very concrete,* and there is violence in the act.

*Black Is . . . Black Ain't,* filmmaker Marlon Riggs insists—quoting the popular words of a character from Ralph Ellison's classic novel *Invisible Man*—yet the purpose of the masks which, like a cross, Africans in the West are forced to wear, is to erase the ineluctable contained in that expression and in its place erect borders and signs of convenient definition. By making such role-playing a requisite of survival, by insisting on what has come to be known as *the fact of blackness,* society also makes those who are forced into the roles complicit in their own self-otherization and self-negation, in the violation through denial of their own true selves.

Shoga's installation and her display cabinet of identities speak to the general experience of Africans in diaspora, but even more specifically to the unique experience of those who are descended from across the races. It addresses not only the enormous political symbolism of physiognomy, but also the psychological repercussions of racial insistence on impossible clarity. *Like Something Not Very Concrete* reminds us of Fanon's explorations into the psychology of blackness in a racial culture,[16] and of the implications of social conventions that question the individual's claims to basic humanity beyond ill-determined specifics and certainties. The protagonists in the video question their sense of self as well as society's intervention in the definition of this sense of self; their encounters with the pressures of this intervention become a test for their own resilience and a process of self-discovery and affirmation without which wholeness is denied them. In the end, their ritual of masking and unmasking, which is also manifest in their ever-changing hairstyles,[17] concludes in another ritual. This ultimate ritual is quite analogous to the powerful rhetoric inscribed in the image of a naked Marlon Riggs running through thickets in his last work, *Black Is . . . Black Ain't,*[18] stripped of all masks, searching now for that essence which none other must venture to confine or define. In their search Shoga's characters reaffirm the essential indeterminancy of identity, like something not very concrete, a fluidity of history, circumstance, and choice, without whose tangles we are deprived and unwhole.

Africa's contemporary artists direct us to important issues and questions that engage them as well as their cultures of location and origin, as is evident from the work of these artists. More importantly, they challenge us to challenge ourselves, to cultivate and preserve those corridors of crossing through which our lives are enriched, our experiences broadened, and our beings ennobled.

## NOTES

1. Octavio Zaya and Anders Michelson, *Interzones: A Work in Progress* (Copenhagen: Tabapress, 1996), 13.

2. Salman Rushdie, *Imaginary Homelands: Essays and Criticism 1981–1991* (London: Granta Books, 1991).

3. Christopher Okigbo, "Distances," from *Labyrinths: Poems* (Oxford: Heinemann Educational Books, 1971), 60.

4. It has been speculated that the Xanadu of Coleridge's unfinished poem "Kubla Khan (A Vision in a Dream)," written in 1797 and published in *Christabel and Other Poems* (1816), might be Florida, although the only two characters mentioned in the poem, the protagonist Kubla and an "Abyssinian maid" suggest otherwise. Also in the poem, the sacred river Alph runs down "to a sunless sea," a very unlikely scene indeed for Florida.

5. *Spurious Artifacts* (1991), an on-site installation and work in progress.

6. The ultimate migration of the sign in this instance, of course, is the eventual commodification of the Mandela saint, whose face now graces expensive Swiss wristwatches and gold tieclips, very much in the tradition of Ché Guevara's transition from loathed guerrilla leader to Coca-Cola billboard icon.

7. Rian Malan, *My Traitor's Heart* (New York: Atlantic Monthly Press, 1990).

8. Lubaina Himid, *Beach House* (Wrexham, England: Wrexham Library Art Centre, 1995).

9. Toni Morrison, *Song of Solomon* (New York: Random House, 1977).

10. In his excellent essay on Niati's work, Salah Hassan draws a particularly important analogy between Niati's interpretation of Delacroix and Algerian writer and filmmaker Asia Djebar's treatment of the same theme in her novel *Women of Algiers in Their Apartment* (1992) in which, as Hassan points out, she portrays Delacroix as "a thief, a spy and an intruder . . . (whose) painting is viewed as an agent of oppression." See Salah M. Hassan, "The Installations of Houria Niati," *Nka: Journal of Contemporary African Art* 3 (1995): 50–55. For related scholarship see Malek Alloula's seminal and controversial work on the subject, and Edward Said's on the broader questions of Orientalist ideology and its cultural manifestations.

11. Ibid., 53.

12. For my earliest reflections on these questions, see Olu Oguibe, *. . . unbind me* (Lagos: Didi Museum, 1988), and *Statements* (Lagos: Syrian Club, 1989).

13. Sigmund Freud, *Five Lectures on Psychoanalysis,* trans. and ed. James Strachey (Middlesex: Penguin Books, 1957), 39.

14. Paul Celan, "Todesfuge," trans. in Michael Hamburger, *Poems of Paul Celan* (New York: Perasa Books, 1972) and more recently in John Felstiner, *Paul Celan: Poet, Survivor, Jew* (New Haven: Yale University Press, 1995).

15. Geoff Dunlop, dir., *Jean-Michel Basquiat: Shooting Star* (BBC Channel Four, 1990).

16. Especially in Frantz Fanon, *Black Skin, White Masks* (London: Pluto Press, 1986).

17. To understand fully the place of hair in African diaspora identity (and thus its role in Shoga's installation) see Lisa Jones, *Bulletproof Diva: Tales of Race, Sex, and Hair* (New York: Penguin Books, 1995).

18. Marlon Riggs, dir., *Black Is . . . Black Ain't* (1995).

OKWUI ENWEZOR

# Between Localism and Worldliness

*Daughters of the Dust,* the skillful and imaginative work by filmmaker Julie Dash, is a film based on the epic tale of dislocation, migration, and acts of cultural survival forged under the despotic and inscrutable memory of slavery. Executed with an economy of narrative intensity, the film makes clear through a series of *mise-en-scènes,* flashbacks, and cultural constructions that once a paradise is lost it is never regained. Far from relying on the abstractions and exaggerated sense of selfhood that often accompany and sometimes disfigure the narration of the social crisis many exiled communities experience during moments of transition, in-betweenness, and loss, Dash's film meditates on questions of memory, love, kinship, faith, apostasy, hybridity, and political and cultural agency. In the film, explorations of individual freedom and desire are posed against communal, racial, and family cohesion, with metaphor to conjure and wrestle with the demons of displacement. The film is built around the structure of the microscopic social and cultural text, through the minutiae of domestic and personal narratives sited within the grand territory of history. Though she did not address explicitly the issue of return, it remains clear throughout that one of the structural concepts of the film relies on how to imagine one's identity in the temporal shift of modernity, that is, how to broach the unspeakable terror and the near impossibility of return once one's consciousness is propelled forward by that idée fixe of modernity known as progress. Yet the film vividly remains one that attempts to articulate the story and the beleaguered temporality of the African presence in the New World.

I want to use this sense of beleagueredness to touch on the recent debates raised by Ebonics. As many people see it, Ebonics is part of the ongoing fraught legacies of slavery and racism, the uneven mixture of African and European cultural norms. This hybrid process produced what today many refer to as a "black English," raising questions about linguistic otherness. However, the question is not so much whether Ebonics can be pegged into a valid pedagogical category as a language of choice for African-American students who are said to be deficient in standard English, hence their poor academic performance; the question that is more immediate for me is whether it is a valid vehicle for sociocultural expression. A very cursory survey of hip-hop makes such questions unnecessary. One could certainly affirm that Ebonics is as valid for African-Americans as Latin is to the liturgical language of the Vatican; it is a tool for making visible expressions of desire, selfhood, identity, etc. . . .

Listening to the various voices in Dash's film, particularly the tone and texture of their "English" and how they bend and render certain "English" phrases into new systems of meaning and creative discourse, provides another key into how we listen to the Ebonics debate. Whether it is a subcultural argot or deformed language, such deep-seated ideological and political meditation on culture, language, and translation locates the pedagogues of identity and cultural politics in the thick of the struggle over issues of race, selfhood, community, and deep into the recesses of cultural expressions and social freedom. That this issue would reemerge during the giddy celebration that characterizes globalization offers us an opportunity to look more searchingly at the sort of narratives that accompany what we consider global culture, especially if it is a global culture that is inattentive to the particularities of alternative modes of expression clustered within many sections of the world. My point here isn't to compose a praise song to Ebonics, but then neither am I interested in demonizing it. The fluid positions from which *Daughters of the Dust* and Ebonics emerge say something about the notion of translation and turn language not just into a site of contestation but one that constantly suffers revision as we follow the tangled routes of the movement of populations, cultures, and peoples from one part of the world to another.

On this account, I am as interested in exploring how language spoken in the disadvantaged locales of displaced communities can be bent to new uses as I am fascinated by the anxiety it produces in so-called settled communities, and the way established categories of meaning are trangressed in those seemingly settled parts of the global village. It would seem to me, if we consider further the Ebonics question, that the main issue isn't the students who have thus far failed to fully assimilate into the settled community (partly because of their sociocultural background and partly because, as someone put it, they have been too hooked on Ebonics) but the very nature of what we call English today. Quite acutely, we can see that colonization works in mysterious ways; in this case, what the English language seems to have suffered is a revision. Such a revision contests any essential notion that language, along with culture and identity, is a fossilized object of social and cultural discourse. As can be seen in this debate, any encounter, however deeply rooted in systems of oppression, violence, and cultural erasure, places itself at the service of an inevitable and unsociable dissolution of a prior order; it fixes us in the emergence of a new temporality. In this case the revisionary portents of diasporicity become equally an analysis of the ideological crisis of the colonial language. It opens up new paths for us to experience how postcolonial procedures have inscribed themselves in the discursive sites of global politics and culture.

But let me return to *Daughters of the Dust* as a particularly poignant moment around which one may experience the objectification and limits of a colonial language through the mass cultural medium of film. As we watch characters and listen to their voices, we hear clearly through the different accents and ways of employing the English language how the complex weave of tongues has been seasoned by movement. Diasporicity, fragmentation, dislocation, reterritorialization, exile, etc., all these motifs of transnational human culture and fraught modernity collude to make new meanings of identity. They produce new cultural possibilities out of the scatological travelogue of twentieth-century migrancy.

Beyond languages and accents, this film also illustrates other points. Told on the eve of the migration of the Peasant family from a place called Ibo Landing, in the Sea Islands of Georgia, in the United States, their migration (that leaving behind that is less a movement of progress than a wound) is an exemplary text on that form of scattering and separation which bears portentous similarity to the context of the African artists I discuss in this essay. The presence of these artists in many parts of the world captures in manifold ways that reality of arrivals and departures that every day is played out in many airport terminals, train stations, docks, etc., throughout the world. How has the language of these artists changed since migration? How have their identity, sense of placelessness, or presence been altered by re/dislocation? And how have their identity, placelessness, and presence affected, altered, or extended the normative forms of expression in the sites they occupy? One question remains very clear: does migration necessarily mean the leaving behind of one's own country, culture, and ethnic enclave, or does it involve other forms of traveling that mean more than the physical transversal of borders?

This last question is important if we consider how much things have changed in the last half-century. The movement of populations—from rural to urban, agrarian to industrialized, village to cosmopolitan, national to postnational and transnational—provides keys to new articulations about the meanings of identity, identification, affiliation, allegiance, alliance, etc. Introducing concepts of hybridity, ambivalence, and indeterminacy into the lingua franca of cultural and political discourse, these movements pry open routes into the values of ethnicity, origin, and authenticity. Furthermore, such reroutings not only question, they also unsettle allegiances and make clear sites of myriad political, cultural, social, and expressive thought, so that speaking of "black" Africa has become not only an inadequate point of classification and differentiation, but anachronistic. In this regard, various discourses are beginning to recognize the validity of Maghrebian, Caucasian, and other histories as integral to the ways we define and expand the notion of who and what is African.

Thus in speaking of Africa today—given its complex colonial and postcolonial history—we need to ask the obvious question: how do the struggles of independence, the problems of the national sovereign state, the expanding definition of national culture, citizenship, and cosmopolitanism, which are partially linked to economic malaise, social obsolescence, and political destabilization, define subjectivity for us? What role does the notion of individual freedom and desire play in the construction of identity? How do such definitions provide affective processes of critical thinking, radical revision, translation, postnationality, etc.? In terms of examining each procedure, what are the critical tools available for the work of the cultural historian? These questions come with a particular urgency, for in the peculiar circumstances of globalization, however parochial its current positions, we have to account for new diasporic formations that have become part of the postcolonial experience of African artists and intellectuals. We need to investigate and analyze the cultural and intellectual productions that are based on this experience of diaspora, to explore how the conditions of exile and expatriation provide new motifs and challenges to the new discourse of Africa in the late twentieth century. This investigation should contend not only with exile and expatriation in relation to movement into the crowded metropolises of the western hemisphere, but also with the transnational movements we see occurring today in Lagos, Abidjan, Johannesburg, Dakar, and Cairo.

While movement and migration (forced or otherwise) have been perpetual motifs of the twentieth century, and while many of us often fox on the problematics of dislocation and displacement, particularly those made through spatial distinctions, between here and there, home and exile, we mustn't forget that vast numbers of migrations happen internally, within bounded national territories. Even when these movements happen internally, they are not always predicated on a shattered spatiality as we find in certain segments of say, Johannesburg, where particular kinds of migrants form clusters. These clusters, in addition to redefining the spatial character of the city, bring to those sites new cultural archetypes, languages, etc., which often enter into competition with the rooted, settled communities.

Though these convergences often serve as real metaphors for conflict, it is the realization of a new temporality within the spatial problematics of the city that makes the process interesting. Thus it is possible to live in one's own country, city, and culture, and remain as distinctly alienated and distant from its social procedures as those who journey out to the strange beyond of the global metropolis; this minimally recognized condition—composed of migrancy, placelessness, exile, and displacement—for me serves as a real metaphor for what today's contemporary African artist embodies. They travel both at home and abroad; they travel physically and psychically; they migrate inbetween the pixelated and information-saturated sites of the cyberworld; they inhabit the complex matrices of popular culture (electronic, digital, and aural virtualscapes) that form part of the transterritorial dimension of the global network and exchange systems. Additionally, each of these artists brings very disparate attitudes and experiences to the zones where they trade (not only in symbolic exchanges), and they help to redefine and reshape the contours of contemporary cultural practice.

These habitués engage in critical conversation with the thorny issues of place, identity, and memory, which acquire new meaning insofar as they detotalize and deconstruct a performative African psychic space from a homogenized, political economy of race and authenticity to one of multiple identities. In evoking some of the problematic terms that have served as canon fodder for African identity discourse, I have no interest in repeating here all those worn-out modes of postcolonial address that insistently and invidiously set up binary distinctions within the practices of those African artists who live on the continent and those who don't. Nor am I interested in the other distinctions that separate their practices into native/foreign-born, the authentic and inauthentic; in such echo chambers nationalism, ethnocentrism, and racism circulate their noxious fumes. Still, it would be fair to concede the point that indeed, certain differences do layer some of the above distinctions. However, that would be an obvious point to make, and singling out differences is only helpful insofar as we understand that we can only deploy them to establish paradigmatic attitudes that exist between disciplines, discourses, locations, and practices. In so doing my argument rests on what I perceive to be one of historical necessity.

But let us return once more to how those zones in which the strategic signs of African contemporaneity circulate and what they are bounded by. I think of my own context of transnationalization: as an Igbo, Nigerian-born, New York–based, Johannesburg resident, world citizen, and latterly American citizen, moving in and out of identities and places, it seems I have developed a certain partiality to those kinds of cultural positions that are signed by contingency, indeterminacy, and cosmopolitanism that are thrilling in the new kinds of contemporary practices I find in the work of African artists at home and abroad. Their names are varied: Pascale Marthine Tayou, Georges Adeagbo, Moshekwa Langa, Jo Ratcliff, Bili Bidjocka, Folake Shoga, Wageshi Mutu, Mary Evans, Olu Oguibe, Kendell Geers, Kay Hassan, Oladélé Bamgboyé, Bodys Isek Kingelez, Yinka Shonibare, Ouattara, Ike Ude, Donald Odita, Abdoulaye Konate, Peet Pienaar, Tracey Rose, etc. They come from Johannesburg, Douala, Lagos, Cotonou, Kinshasa, Cape Town, Aba, Nairobi, Bamako, and Dakar, and they live and work in New York, London, Amsterdam, Paris, Berlin, and Tokyo.

Each of these artists (there are in fact too many to list) works within many critical, avant-garde procedures: video, film, installation, actions, performances, photography, and digital technology. Though they do not eschew traditional means of making art, in confronting their works I am forced to recognize them as being deeply located within conceptual and postmodernist matrices. They speak to me in a manner that I will call strategic performances. Often based on the critique of various hegemonic social, political, and cultural practices, their works aim to destabilize, challenge, and reinscribe the global terrain of culture as a complex territory in which they can stake a claim as much as can any contemporary practice coming out of England, the United States, France, or Germany. In so doing they invoke and highlight for me many transgressive and subversive movements and critical and radical articulations: for example, Shoga's investigations of gender and hybridity; Bamgboyé's exploration of place and memory; Langa's and Rotimi Fani-Kayode's interrogation of otherness and sexual difference; Oguibe's critique of imperialism; Bidjocka's flâneurship; Ude's theatricalization of masculinity and performative inversion of gender codes; Geers's subversive conceptual and politicized gestures; Kingelez's utopian and postmodern vision of architecture and the African city; and Adeagbo's linguistic puns, taxonomy, and commodity fetishism.

Nonetheless, talking about these artists in this way does not give us a full picture of what it is to be an African artist in the age of globalization; all it says and repeats is what has always been known, i.e., that Africa is made up of multiple and disparate identities, cultures, territories. However, in following this most traditional of analytic routes, what we are given is a set of imperatives with which to examine what Homi Bhabha has termed "measure of dwelling," which speaks to how we define and determine belonging and place. What, in the late twentieth century, are the conditions of belonging and dwelling for Africans who live in places other than Africa? Let us examine briefly two concepts: the *jus solis* and *jus sanguinis*. *Jus solis* simply means belonging determined by place and the length of sojourn therein, and *jus sanguinis* is belonging based on blood, ancestry, and natal origin, conceptions of identity determined through biological, political, cultural, and territorial differences. In Cuba the popular revolutionary slogan announced this position quite forcefully: *patria o muerte* (fatherland or death). In the context of current local/global tensions, the nationalism inherent in such sloganeering seems not so much quaint as outdated.

For if belonging is circulated and secured through parricide and the atavistic attachment to land or place, as the Cuban slogan and places such as Bosnia and Rwanda seem to suggest, what happens when the reality of displacement or exile challenges the craggy edifices of these identificatory systems? To whom does the fatherland belong? The migrant or guest worker with minimal political rights, without claim by blood or the tenure of land? I think not. The reason I have repeatedly raised these issues is that very few of the discussions about belonging, identity, nationality, and exile have adequately explored that hinge where many Africans dwell, think, create, or in some cases, while away their time dreaming paradoxically of home. And where exactly is home for these people? And where home has become unimaginable except in frayed old black-and-white photographs, what set of imperatives within the nascent narratives of crossing, settling, dwelling, and transterritorialization do such immigrants conjure up to locate themselves in the new land and to stitch the unruly patterns of home? How do they accommodate the locations of departure and arrival? What services those journeys, whether voluntary or coerced?

The reader can very well imagine that there are very few answers for such fraught questions, particularly as we focus on those expressive and social languages acquired in the temporal fissures of postcolonial migration. Making sense of the new temporalities and spatial configurations many Africans have entered into today—between what James Clifford calls localism and worldliness—cannot happen until we learn to acquaint ourselves intimately with how African subjectivity has been defined within this reality. Moments of migration forged by contingent histories repeat through various signs, little bits and pieces of remembered conversations (articulations predicated on the falling away, the fragmentation of collective memory). Even while attempting to disguise those accents of foreignness and hybridity—that exemplary piecing together, part myth and part experience—one easily slips between different forms of speech, from colloquial Igbo to clipped British stiff-upper-lip English; from Lingala to Ivorien French; from Angolan Portuguese to Liberian Creole to Nigerian pidgin; from classical Yoruba to Swahili to Arabic and Hindi. All these, at any rate, describe Africa, at home and abroad.

In such a moment of contingent, indeterminate cultural histories, one would have to question why few discourses of contemporary Africa have taken notice of these patterns and textures of migration and movement. There are reasons for this: First, the degree to which the patterns of migration have been figured seems always to narrate that liminal space as only a temporary one; hence, the need to define and forge a broader affiliation of Africans in a foreign place (which diaspora suggests) was seen as unnecessary, for one always returns. Secondly, many African immigrants who left before or during the period of independence never imagined that various political and economic emergencies could so easily attenuate the desire to return. Moreover, many found secure situations for themselves in the places they had settled and so those places became legitimate homes away from home.

On a different level, many cultural historians still fixed in an absolutist demarcation of cultural and ethnic boundaries, and those following anthropological and ethnographical practices, persist in ignoring the cosmopolitan character of places such as Dakar, Lagos, Accra, and Abidjan. These cities have, from the early part of the twentieth century, accommodated a vibrant mixture of African cultures. Many of the most memorable moments of this cosmopolitanism can be found in the enduring portraits of the citizens of these capacious cities from such photographers as Mama Casset, Mix Gueye, Meissa Gaye, Seydou Keita, Ricardo Rangel, Peter Magubane, Alf Khumalo, and Sala Casset, who were working in the colonial port city of Saint Louis in Senegal, and in Dakar, Maputo, Johannesburg, and Bamako from the 1920s to the present.

While these vibrant mixtures were happening, creating allegorical layerings and collages of modernity in the process, another group of African artists such as Ben Enweonwu, Uzo Egonu, Gerard Sekoto, Iba Ndiaye, Skunder Boghosian, El Salahi, and Ernest Mancoba were already working in such places as London, Paris, and New York. These artists, conversant with the political movements of independence and the social struggles of their home countries, wove into their critical practices questions ranging from postcolonialism to pan-Africanism. In the specific cases of Egonu and Sekoto, both of whom had highly productive careers in England and France, respectively, the postcolonial articulation of diaspora, displacement, ambivalence, and contingency finds its most critical voice. In the work of these two artists, both of whom barely returned to their home countries (Nigeria for

Egonu and South Africa for Sekoto), one finds the persistence of memory in the exuberantly (and at times darkly) limned aspects of life in the home country. They paint a vision of the world where the inward gaze explodes and looks outside, creating visual and psychological moments filled with desire, irresolution, ennui, and disjunctive spatialities that detonate like time bombs in the strange weather of European modernist art, which has accorded them few acknowledgments.

If one pursues a more detailed picture of contemporary Africa in the late twentieth century, one finds that diaspora is not an equivocal term that excludes African artists. In its incremental and divergent formations, thinking about the links that diasporicity offers as tools to critically appraise the art of the continent does indeed give us access to the transcontinental and transnational regimes of the contemporary production of her artists. It short-circuits any essentialist reading of "African" as embedded in a timeless warp of precolonial African traditions. James Clifford put it nicely when he wrote that diaspora could be "seen as potential subversions of nationality—ways of sustaining connections with more than one place while practicing nonabsolutist forms of citizenship." It seems to me that the terms under which we negotiate the sense of what is African in the late twentieth century is predicated on the values of this nonabsolutist, nonessential form of affiliation. This is what the artists gathered under the rubric of this exhibition make clear: theirs is a world that is not circumscribed by any absolutist identity or territory. In this sense the exhibition raises key questions not only for the audience that will have its views of what and who an African artist is, but equally, those critical Western establishments that will no doubt continue to attempt to sequester these artists into disqualified ethnic categories. But more importantly, the artists pose another more salient and lasting question: like the Peasant family in *Daughters of the Dust,* as we cross and settle and resettle, how do the new accents we acquire during the course of migration or contact with other cultures change our positions of affiliation? How do we secure new communities, embody diverse identities (sexual, gendered, racial), reterritorialize vestiges of the cultures of home, experiment with new ways of being and making, and create new economies of exchange and circulation for stories and symbolic and political values? Indeed, what are the ways one is and becomes African in the surging tumult and noise of the millennial clamor for a homogenized—and commodified—global identity?

**Oladélé Bamgboyé**
Photo, installation, and video artist. Born in 1963 in Benin, Nigeria. Grew up and studied in Glasgow, Scotland. Has worked in Canada and Germany; currently lives and works in London. Has had solo exhibitions in Scotland, Estonia, and Germany. Has participated in major group exhibitions in Asia, Europe, and North America, including the "Sorak International Art Biennial" in South Korea; "Prospect '96" in Frankfurt; "Inclusion/Exclusion: Art in the Age of Post-Colonialism and Global Migration" in Graz, Austria; "Self-Evident" at the Ikon Gallery in Birmingham, England; and "In/Sight: African Photographers, 1940 to the Present" at the Solomon R. Guggenheim Museum in New York. Winner of numerous awards, including grants from the Scottish and British Arts Councils as well as the Gulbenkian Foundation New Horizons Bursary for New Ideas in Filmmaking.

**Bili Bidjocka**
Painter and installation artist. Born in 1962 in Douala, Cameroon. Moved to France at age twelve. Currently lives and works in Paris. Has had solo exhibitions in Europe and North America and has participated in several group exhibitions, most recently the "2nd Johannesburg Biennale" in South Africa; "Interzones" at the Kunstforeningen, Copenhagen; and "Otro País: Escalas Africanas," which traveled throughout Europe.

**Gordon Bleach**
Photo and installation artist. Born in 1953 in Zimbabwe. Holds a doctoral degree in applied mathematics from the University of Cape Town in addition to a master's degree in photography from the Rochester Institute of Technology. Currently lives and works in Gainesville, Florida. Has shown extensively in South Africa and recently participated in the exhibition "In/Sight: African Photographers, 1940 to the Present" at the Solomon R. Guggenheim Museum in New York. Winner of various awards, including a Samuel H. Kress Travel Fellowship and a Federal Travel Grant from the Center for African Studies at the University of Florida.

**Kendell Geers**
Conceptual and installation artist. Born in 1968 in Johannesburg, South Africa. Currently lives and works in Johannesburg. Has had solo exhibitions in Johannesburg and Durban, South Africa, and taken part in numerous group exhibitions in Africa, Europe, and the Americas. Exhibited at the Aperto of the "XLVI Venice Biennale," as well as the "2nd Johannesburg Biennale" and the "5th Bienal of Havana."

**Lubaina Himid**
Painter and installation artist. Born in 1954 in Zanzibar, Tanzania. Holds degrees from the Wimbledon School of Art and the Royal College of Art in London, and is currently pursuing a doctorate at the University of East Anglia. Director of the Elbow Room, an alternative space in London, between 1986 and 1990. Has had several solo exhibitions in Europe and the United States. Has taken part in major group exhibitions such as the "5th Bienal of Havana" and "The Other Story" at the Hayward Gallery, London.

**Marcia Kure**
Painter and printmaker. Born in 1970 in Kano, Nigeria. Studied at the University of Nigeria in Nsukka. Currently lives and works in Nsukka. Has had solo exhibitions in Germany and and Nigeria and taken part in group exhibitions in Europe, Africa, and North America, most recently "Oshogbo-Nsukka" in Gersthofen, Germany, and "Fragments and Songs" in Lagos, Nigeria.

**Houria Niati**

Painter, singer, installation artist. Born in 1948 in Khemis-Miliana, near Algiers. Studied in Algiers and England. Has traveled throughout Europe and currently lives and works in London. Has had several solo exhibitions and has participated in many group exhibitions in Europe, the Middle East, and North America, including the "Third International Biennial" in Sharjah, United Arab Emirates; "Contemporary African Women Artists" at Cornell University in Ithaca, New York; and "New Visions: Recent Works by Six African Artists" at the Zora Neale Hurston National Museum of Fine Arts in Eatonville, Florida.

**Olu Oguibe**

Conceptual artist, poet, critic, curator. Born in 1964 in Aba, Nigeria. Studied in Nigeria and at the University of London, where he received a doctorate in art history. Currently Stuart S. Golding Chair in African Art at the University of South Florida. Has had many solo exhibitions in Africa, Europe, and Australia, and has participated in group exhibitions in Africa, Europe, and North America, including the "2nd Johannesburg Biennale" in South Africa; "Seven Stories about Modern Art in Africa" at the Whitechapel Gallery, London; and "La Città Idéale" at the Museo Domestico di Pettineo, Italy. Winner of several awards, including the Nigerian National Council for Arts and Culture Prize, the Arts Council of Great Britain New Art Collaborations Travelling Exhibitions Grant, and the 1992 Christopher Okigbo All-Africa Prize for Literature. Editor of *Nka: Journal of Contemporary African Art* (New York) and board member of *Third Text* (London); has published and lectured extensively on African art.

**Tracey Rose**

Video and installation artist. Born in 1974 in Johannesburg, South Africa. Received a bachelor's degree from the University of Witwatersrand in South Africa. Currently lives and works in Johannesburg. Has participated in several group exhibitions, including the "2nd Johannesburg Biennale"; "Hitchhiker" at the Generator Art Space, Johannesburg; and "Purity and Danger" at the Gertrude Posel Gallery, Johannesburg.

**Folake Shoga**

Video and installation artist, filmmaker. Born in 1955 in Ibadan, Nigeria. Studied at University of Newscastle, Newcastle-upon-Tyne. Currently lives and works in Bristol, England. Has had numerous solo exhibitions and taken part in many group exhibitions such as "Four x 4"; "Let the Canvas Come to Light With Dark Faces"; and "Reclaiming the Madonna." Winner of several awards, including the Arts Council of Great Britain Bursary for Film and Video; the Ganier of South West Arts Award; the HTV/SWA Experimental Film Awards; and grants from the Arts Council of Great Britain and the Gulbenkian Foundation.

CHECKLIST

1 Oladélé Bamgboyé
Panel #4, from the series "Das Lichthaus"
1989
photo installation
36 x 48 inches

2 Oladélé Bamgboyé
Images 1, 3, 5, 6 from the series "Celebrate"
1994
photo installation
35 x 45 inches

3 Bili Bidjocka
*Untitled (Witches' Ball)*
1994
mixed-media installation
dimensions variable

4 Bili Bidjocka
*Explicit Lyrics*
1995
mixed-media installation
approx. 80 x 192 inches

5* Gordon Bleach
*Backslash*
1997
mixed-media installation (bricks and glass)
approx. 240 x 60 inches

6 Gordon Bleach
*Blue Mote: xanadu, Florida* from "Double Vision: The Mote Series"
1997
computer-generated photographic print
72 x 36 inches

7 Gordon Bleach
*Gold Mote: gomo, Zimbabwe* from "Double Vision: The Mote Series"
1997
computer-generated photographic print
72 x 36 inches

8 Gordon Bleach
*Red Mote: rover, Mars* from "Double Vision: The Mote Series"
1997
computer-generated photographic print
72 x 36 inches

9 Kendell Geers
*Title Withheld (Untitled)*
1993
installation
dimensions variable

10 Kendell Geers
*Title Withheld (CROSS/ING)*
1997
intervention (crate installation)
dimensions variable

11 Kendell Geers
*Title Withheld (Out of Africa)*
1997
situation
dimensions variable

12 Lubaina Himid
*Restoring the Balance*
1987
acrylic on wood, 3 woodcuts
72 x 48 inches

13 Lubaina Himid
*African Garden: Billie Holiday, Mashulan, Augusta Savage*
1993
triptych: acrylic on canvas with mixed media
approx. 84 x 144 x 12 inches

14* Marcia Kure
*African Woman*
1996
ink, pencil, and charcoal on paper
48 x 35½ inches

15 Marcia Kure
*Woman at a Wedding*
1996
ink, pencil, charcoal, and pastel on paper
48 x 35½ inches

16* Marcia Kure
*Woman Weeding*
1996
ink, pencil, and charcoal on paper
48 x 35½ inches

17 Marcia Kure
*Woman as Akuaba*
1997
ink, pencil, charcoal, and pastel on paper
48 x 35½ inches

18* Marcia Kure
*Woman as Dancer*
1997
ink, pencil, charcoal, and kolanut pigment on paper
48 x 35½ inches

19 Marcia Kure
*Woman as Warrior*
1997
ink, pencil, charcoal, and pastel on paper
48 x 35½ inches

20 Marcia Kure
*Woman Giving Birth*
1997
ink, pencil, charcoal, pastel, watercolor, and kolanut pigment on paper
48 x 35½ inches

21* Marcia Kure
*Woman in Purdah I*
1997
ink, pencil, charcoal, and pastel on paper
48 x 35½ inches

22 Marcia Kure
*Woman in Purdah II*
1997
ink, pencil, and charcoal on paper
48 x 35½ inches

23* Marcia Kure
*Woman in Purdah III*
1997
ink, pencil, charcoal, and pastel on paper
48 x 35½ inches

24 Marcia Kure
*Woman Suckling Her Male Child*
1996
ink, pencil, and charcoal on paper
48 x 35½ inches

**25*** Houria Niati
*No to Torture*
1982–83
mixed-media installation
dimensions variable

**26** Houria Niati
*Bringing Water from the Fountain Has Nothing Romantic about It*
1991
mixed-media installation
dimensions variable

**27*** Olu Oguibe
*Mementos*
1996
multimedia installation
dimensions variable

**28** Olu Oguibe
*Buggy (For an Unknown Child)*
1997
multimedia installation
dimensions variable

**29*** Olu Oguibe
*Memorial to a Youth*
1997
multimedia installation
dimensions variable

**30** Tracey Rose
*Untitled 1940—1996*
1997
birth certificates
dimensions variable

**31** Tracey Rose
*Untitled*
1997
video installation
dimensions variable

**32** Folake Shoga
*Like Something Not Very Concrete*
1994
mixed-media installation
dimensions variable

*Indicates works not included in traveling exhibition.

## SELECTED BIBLIOGRAPHY

Compiled by Chika Okeke

### BOOKS


Arnold, Marion I. *Women and Art in South Africa.* New York: St. Martin's Press, 1996.

Beier, Ulli. *Art in Nigeria, 1960.* Cambridge: Cambridge University Press, 1960.

___. *Contemporary Art in Africa.* London: Pall Mall Press, 1968.

Berman, Esmé. *Painting in South Africa.* Johannesburg: Southern Book Publishers, 1993.

Darah, G. G., and Safy Quel, eds. *Bruce Onobrakpeya: The Spirit in Ascent.* Lagos: Ovuomaroro Gallery, 1992.

Fisher, Jean, ed. *Global Visions: Towards a New Internationalism in the Visual Arts.* London: Kala Press/Institute of International Visual Arts, 1994.

Flamand, Alain. *Regard sur la peinture contemporaine du Maroc.* Casablanca: Société d'Édition et de Diffusion Al-Madariss, 1983.

Fosu, Kojo. *Twentieth-Century Art of Africa.* Zaria, Nigeria: Gaskiya Corporation, 1986.

Geers, Kendell, ed. *Contemporary South African Art: The Gencor Collection.* Johannesburg: Jonathan Ball Publishers, 1997.

Guez, Nicole. *Contemporary African Art,* 2nd ed. Paris: Association Afrique en Création, 1996.

Hultén, C. O., et al. *Moderne Konst i Afrika.* Lund, Sweden: Kalejdoskop, 1978.

Jegede, Dele. *Trends in Contemporary Nigerian Art: A Historical Analysis.* Ann Arbor, Mich.: University Microfilms International, 1983.

Karnouk, Liliane. *Contemporary Egyptian Art.* Cairo: American University in Cairo Press, 1995.

___. *Modern Egyptian Art: The Emergence of a National Style.* Cairo: American University in Cairo Press, 1988.

Kelly, Bernice M. *Nigerian Artists: A Who's Who and Bibliography.* Ed. Janet L. Stanley. London and New York: National Museum of African Art, Smithsonian Institution Libraries, Washington, D.C., by Hans Zell, 1993.

Kennedy, Jean. *New Currents, Ancient Rivers: Contemporary African Artists in a Generation of Change.* Washington, D.C.: Smithsonian Institution Press, 1992.

Konaté, Yacouba. *Christian Lattier: Le sculpteur aux Mains Nues.* Saint-Maur, France: Sepia, 1993.

Lucie-Smith, Edward. *Race, Sex, and Gender In Contemporary Art.* New York: Harry N. Abrams, 1994.

Manaka, Matsemela, ed. *Echoes of African Art: A Century of Art in South Africa.* Braamfontein: Skotaville, 1987.

Mount, Marshall Ward. *African Art: The Years Since 1920.* 1973. Reprint, New York: Da Capo Press, 1989.

Oguibe, Olu. *Uzo Egonu: An African Artist in the West.* London: Kala Press, 1995.

Oguibe, Olu, ed. *Third Text* (London) Africa Special Issue, 23 (1993).

Ottenberg, Simon. *New Traditions from Nigeria: Seven Artists of the Nsukka Group.* Washington, D.C. and London: Smithsonian Institution Press and National Museum of African Art, 1997.

Powell, Richard J. *Black Art and Culture in the Twentieth Century.* New York: Thames and Hudson, 1997.

Sealy, Mark, and Jean Loup Pivin, eds. *Rotimi Fani-Kayode and Alex Hirst.* Paris: Editions Revue Noire and Autograph, 1996.

Tadesse, Taye. *Short Biographies of Some Ethiopian Artists.* Addis Ababa: Kuraz Publishing Agency, 1991.

Williamson, Sue. *Resistance Art in South Africa.* Cape Town: David Philip, 1989.

Williamson, Sue, and Ashraf Jamal. *Art in South Africa: The Future Present.* Cape Town: David Philip, 1996.

### EXHIBITION CATALOGUES

*Africus: Johannesburg Biennale, 20 February–30 April 1995.* Johannesburg: Greater Johannesburg Transitional Metropolitan Council, 1995.

Araeen, Rasheed. *The Essential Black Art.* London: Kala Press, 1988.

___. *The Other Story: Afro-Asian Artists in Post-war Britain.* London: South Bank Centre, 1989.

*Art contemporain Arabe: Collection du Musée du L'Institut du Monde Arabe.* Paris: Institut du Monde Arabe, 1988.

*Art/Images in Southern Africa.* Stockholm: Culture House, 1989.

*Bruly Bouabré, Frédéric.* Heidelberg: Edition Braus, 1993.

*Contemporary African Art.* London and New York: Studio International, 1969.

*Contemporary African Artists: Changing Tradition: El Anatsui, Youssouf Bath, Ablade Glover, Tapfuma Gutsa, Rosemary Karuga, Souleymane Keita, Nicholas Mukomberanwa, Henry Munyaradzi, Bruce Onobrakpeya.* New York: Studio Museum in Harlem, 1990.

*Contemporary Art of Senegal.* Baltimore, Md.: Garamond/Pridemark Press, 1980.

*Creative Impulses/Modern Expressions: Four African Artists: Skunder Boghossian, Rashid Diab, Omar Khalil, Amir Nour.* Ithaca, N.Y.: African Studies and Research Center, Institute for African Development, Council for the Creative and Performing Arts, Cornell University, 1993.

Deliss, Clementine, ed. *Seven Stories about Modern Art in Africa.* Paris and New York: Whitechapel, Flammarion, 1995.

Gervasoni, Marie-George. *XLIV Expositione Internazionale d'Arte: La Biennale di Venezia.* Venice: Fabbri Editori, 1990.

Hassan, Salah, ed. *Gendered Visions: The Art of Contemporary African Women.* Trenton, N.J.: Africa World Press Inc., 1997.

*In/Sight: African Photographers, 1940 to the Present.* New York: Solomon R. Guggenheim Museum, 1996.

Levinson, Orde, ed. *I was loneliness: The Complete Graphic Works of John Muafangejo, a Catalogue Raisonné 1968–1987*. Cape Town: Struik Winchester, 1992.

*Lubaina Himid: Revenge*. Lancashire, U.K.: Rochdale Art Gallery, 1992.

*Magiciens de la terre*. Paris: Editions du Pompidou, 1989.

*Malangatana Valente Ngwenya*. Lisbon: Secretaria de Estado da Cultura com a apoio de Embaixada de Moçambique em Portugal e da Sociedade Nacional de Belas-Artes, 1989.

McEvilley, Thomas. *Fusion: West African Artists at the Venice Biennale*. New York: Museum for African Art; Munich: Prestel, 1993.

*Mohammad Omer Khalil: Etchings/Amir I. M. Nour: Sculpture*. Washington, D.C.: National Museum of African Art, 1994.

*Mohammed Kacimi: Paintings, Pastels, Drawings, Events, Writings*. Paris: Editions Revue Noire, 1996.

*New Visions: Recent Works by Six African Artists*. Eatonville, Fl.: The Zora Neale Hurston National Museum of Fine Arts, 1995.

*Otro País: Escalas Africanas* Las Palmas, Spain: Centro Atlántico de Arte Moderno, 1994.

Rankin, Elizabeth D. *Images of Wood: Aspects of the History of Sculpture in Twentieth-Century South Africa*. Johannesburg: Johannesburg Art Gallery, 1989.

*Seen/Unseen*. Liverpool: Blue Coat Gallery, 1994.

*Sixteen Paintings and Prints*. London: Savannah Gallery of Modern African Art, 1994.

*Trade Routes: History, Geography: 2nd Johannesburg Biennale, 1997*. Johannesburg: Greater Johannesburg Metropolitan Council and the Prince Claus Fund for Culture and Development, 1997.

Udechukwu, Obiora. *So Far: Drawings, Paintings, Prints, 1963–1993*. Bayreuth, Germany: Boomerang Press, 1993.

Wijdan, Ali, and Suhail Bisharat, eds. *Contemporary Art from the Islamic World*. London: Scorpion Publishing, 1989.

Zaya, Octavio, and Anders Michelson, eds. *Interzones. A Work in Progress*. Copenhagen: Tabapress, 1996.


ARTICLES


Aniakor, Chike C. "Contemporary Nigerian Artists and Their Traditions." *Black Art* (Jamaica, N.Y.) 4, no. 2 (1980): 40–55.

Araeen, Rasheed. "Our Bauhaus, Others' Mudhouse." *Third Text* (London) 6 (spring 1989): 3–14.

Cousins, Jane. "The Making of Zimbabwean Sculpture." *Third Text* (London) 13 (winter 1991): 31–42.

Enwezor, Okwui. "Redrawing the Boundaries: Towards a New African Art Discourse." *Nka: Journal of Contemporary African Art* (New York), no. 1 (1994): 3–7.

___. "Occupied Territories: Power, Access, and African Art." *Frieze* (London), no. 26 (February 1996): 37–41.

Hassan, Salah M. "The Modernist Experience in African Art: Visual Expression of the Self and Cross-Cultural Aesthetics." *Nka: Journal of Contemporary Art* (New York), no. 2 (1995): 30–33, 72.

Oguibe, Olu. "Africa Explores." *African Arts* (Los Angeles) 26, no. 1 (January 1993): 16–22.

___. "Art, Identity, Boundaries: The Rome Lecture." *Nka: Journal of Contemporary African Art* (New York), no. 3 (1995): 26–33.

Okeke, Chika. "New Painters from Nsukka." *Nka: Journal of Contemporary African Art* (New York), no. 6–7 (1997): 71.

___. "Slashing Wood, Eroding Culture: Conversation with El Anatsui." *Nka: Journal of Contemporary African Art* (New York) 1 (1994): 34–40.

Pearce, Carole. "The Myth of 'Shona Sculpture.'" *Zambezia: The Journal of the University of Zambia* (Harare) 20, no. 2 (1993): 85—107.

Zaya, Octavio. "On Three Counts I Am an Outsider: The Works of Rotimi Fani-Kayode." *Nka: Journal of Contemporary African Art* (New York), no. 4 (1996): 24–29, 72.

PERIODICALS

*ADA: Art, Design, Architecture* (Cape Town). Quarterly.

*African Arts* (Los Angeles). Quarterly.

*Atlántica: Internacional Revista de las Artes* (Las Palmas, Spain). Quarterly.

*Critical Arts: A Journal for Cultural Studies* (Durban, South Africa). Biannual.

*Glendora Review* (Lagos, Nigeria). Quarterly.

*Gallery: The Art Magazine From Gallery Delta* (Harare, Zimbabwe). Quarterly.
*La Revue Noire* (Paris). Quarterly.

*Nka: Journal of Contemporary African Art* (New York). Biannual.

*Third Text: Third World Perspectives on Contemporary Art and Culture* (London). Quarterly.

We are extremely honored to be publishing the catalogue for *Cross/ing: Time.Space.Movement.* This historic exhibition points to the clear realization that, in our shrinking world, there are no longer boundaries in the universe of contemporary ideas. By extension, our problems are also global. And so must be the solutions. Thank you, Dr. Oguibe, for inviting us to be a part of this experience.

TOM PATCHETT
**Track 16 Gallery/Smart Art Press**

**A few of our other special offerings:**

**Dateline Kenya:** The Media Paintings of Joseph Bertiers

**Blood, Sweat & Tears:** Political Art from Northern Ireland, the Spanish Civil War, and the Streets of Los Angeles

**Distant Relations:** Chicano, Irish, Mexican Art and Critical Writing

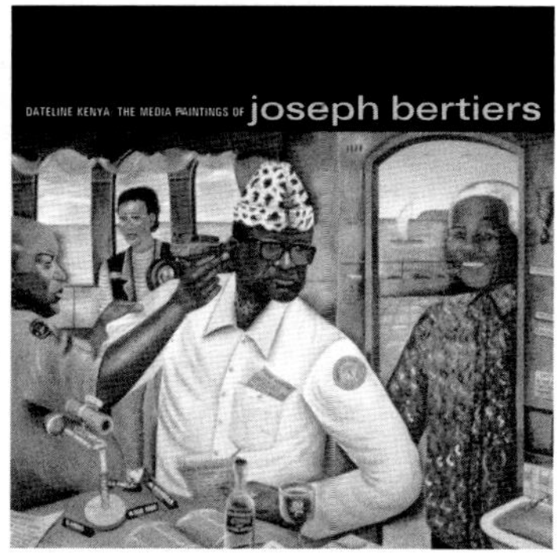


**Dateline Kenya:** The Media Paintings of Joseph Bertiers

Mining CNN and other global media for subject matter, Kenyan artist Joseph Bertiers depicts current events in satirical tableaux which display the artist's formal genius and sly sense of humor. World leaders like Mobutu, Clinton, and Arafat, and celebrities like Mike Tyson and Princess Diana all get the penetrating and witty Bertiers treatment.

Softcover, 9 3/4 x 9 3/4 inches
72 pp
ISBN 1-889195-20-0
$20

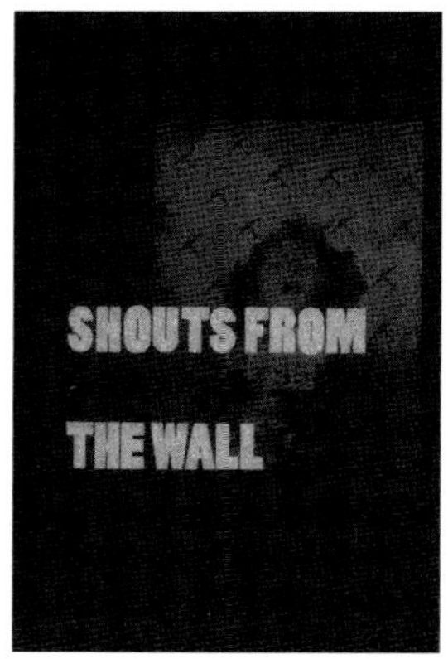


**Blood, Sweat & Tears:** Political Art from Northern Ireland, the Spanish Civil War, and the Streets of Los Angeles

A catalogue in three separate volumes, Blood, Sweat & Tears is the collective title for an unusual trio of exhibitions that looks at the graphics of protest locally and internationally. Three catalogues bound in vellum, introduction by Tom Patchett.

Softcover, 6 x 9 inches
ISBN 1-889195-14-6
(1) Los Angeles: At the Center & On the Edge
(2) No Justice No Peace/Nil Síochán gan Cheart
(3) Shouts from the Wall

112 pp total
The set of 3: $20

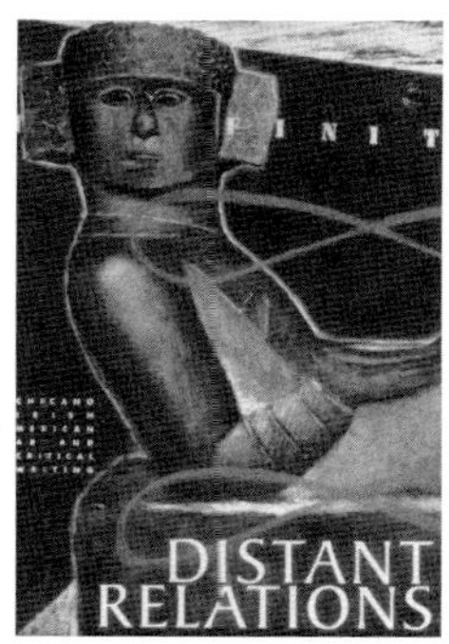


**Distant Relations:** Chicano, Irish, Mexican Art and Critical Writing

Twelve artists and twenty-one authors of Irish, Chicano, and Mexican heritage tackle the issue of how to mount an effective response against dominant cultures in this absorbing anthology of art and critical writing edited by author and curator Trisha Ziff.

Softcover, 6 1/2 x 9 1/4 inches
286 pp
25 color, 34 b&w illustrations
ISBN 0-9646426-1-1
$25